Fictions of Appetite

ITALIAN MODERNITIES

VOL. 17

Edited by
Pierpaolo Antonello and Robert Gordon,
University of Cambridge

PETER LANG
Oxford · Bern · Berlin · Bruxelles · Frankfurt am Main · New York · Wien

Fictions of Appetite

Alimentary Discourses in Italian Modernist Literature

Enrico Cesaretti

PETER LANG
Oxford · Bern · Berlin · Bruxelles · Frankfurt am Main · New York · Wien

Bibliographic information published by Die Deutsche Nationalbibliothek
Die Deutsche Nationalbibliothek lists this publication in the Deutsche Nationalbibliografie; detailed bibliographic data is available on the Internet at http://dnb.d-nb.de.

A catalogue record for this book is available from the British Library.

Library of Congress Control Number: 2013942349

Cover image: Primo Conti, "Fruttivendola" (1915), olio su tela (94 x 138 cm).

ISSN 1662-9108
ISBN 978-30343-0971-4

Hochfeldstrasse 32, CH-3012 Bern, Switzerland
info@peterlang.com, www.peterlang.com, www.peterlang.net

Printed in Germany

Contents

Acknowledgments

This book project grew out of a graduate seminar on "Italian Modernist Literature" that I have been teaching during my years at the University of Virginia. My first thank you, therefore, goes to all the graduate students who attended my course and helped me better define and develop my argument.

My work, at different stages and under different forms, has benefited from the conversations, input, comments and observations of generous colleagues and friends. Their feedback and encouragement is deeply appreciated. The editors of the series "Italian Modernities", Pierpaolo Antonello and Robert Gordon (and the anonymous reviewers) are the latest in a list that should at least mention, among others, Luca Somigli, Ernesto Livorni, John Welle, Paolo Valesio, Marja Harmånmaa, Elisa Biagini and Christina Ball. My additional gratitude goes to Sheila McMillen for her copy-editing and polishing efforts, and Hannah Godfrey at Peter Lang for her help and advice.

The completion of this book was facilitated by three generous awards I received from the University of Virginia: a Sesquicentennial Associateship award (Fall 2008); a PAW (Professors as Writers) award (Fall 2012), and a Research Support Grant in the Arts, Humanities and Social Sciences (Spring 2013). I am thankful to the Dean of the College of Arts and Sciences and the Vice President for Research and Graduate Studies for their consideration and support.

I gratefully acknowledge permission from the journals listed below to reprint, in modified form, some of the following material I have previously published:

"Consuming Texts: Creation and Self-effacement in Palazzeschi and Kafka" originally appeared in *Comparative Literature*, vol. 56, n. 4 (Fall 2004), 300–16.

"Aldo Palazzeschi's *: riflessi* or Writing as Erasure" was originally published in *The Modern Language Review* 99, n. 4 (October 2004), 924–37.

"Dyspepsia as Dystopia? Marinetti's *Le Roi Bombance*" is reprinted with permission from *The Romanic Review* vol. 97, n. 3–4 (May–November 2006), 351–67. Copyright by the Trustees of Columbia University in the City of New York.

"Nutrition as Dissolution: Paola Masino's *Nascita e morte della massaia*" originally appeared in *Quaderni d'Italianistica* vol. 28, n. 2 (2007), 1–20.

"Indigestible Fictions: Hunger, Infanticide and Gender in Paola Masino's 'Fame' and Massimo Bontempelli's *La fame*" originally appeared in *Spunti e Ricerche* vol. 23 (2008), 5–20.

"Outplaying Kronos' Hunger: Massimo Bontempelli's *Gente nel tempo*" originally appeared in *Forum Italicum*, vol. 43, n. 2 (Fall 2009), 385–404.

"Dangerous appetites: Sex and the Inorganic in F.T. Marinetti's Erotic Short-Fiction" originally appeared in *Annali d'Italianistica*, vol. 27, A Century of Futurism 1909–2009 (2009), 139–55.

"Recipes for the Future: Traces of Past Utopias in *The Futurist Cookbook*" originally appeared in *The European Legacy*, eds Pierpaolo Antonello and Marja Harmånmaa, vol. 14, n. 7 (2009), 841–56,

"Milk and Blood: Nursing Capitalism in Luigi Pirandello's *La balia*" originally appeared in *Italian Studies*, vol. 65, n. 1 (February 2010), 33–45.

The image of Primo Conti's "Fruttivendola" (1915), olio su tela (94 x 138 cm) is reproduced by permission of the Fondazione Primo Conti onlus, Fiesole (Firenze, Italy).

I dedicate this book to my wife, Christina, and my daughter, Mila, who sustained me with their love (and other culinary delights) over the years I worked on this project.

Introduction – Fictions of Appetite

La relation à l'écriture, c'est la relation au corps.
— ROLAND BARTHES

Fictions of Appetite: Alimentary Discourses in Italian Modernist Literature investigates the recurrence and aesthetic significance of images of food, appetite, and consumption (and related ones of food rejection, fasting and starvation) in a selection of novels, short stories, poems, essays and plays by Filippo Tommaso Marinetti, Aldo Palazzeschi, Massimo Bontempelli, Paola Masino and Luigi Pirandello that were written in the years from 1905 to 1939 – that is, in decades that many scholars associate with the cultural context of modernism.[1]

Since modernism displayed a new interest in issues of corporeality, in questioning the extensions and limits of bodies and subjects within a new reality marked by increasing scientific and technological discoveries and, as Georges Bataille put it, "human life is still bestially concentrated in the mouth", the liminal site par excellence where words and things are

1 On the peculiar presence/absence of modernism in Italy, see the Introduction to Luca Somigli and Mario Moroni, *Italian Modernism: Italian Culture between Decadentism and Avant-Garde* (Toronto: University of Toronto University Press, 2004). Interestingly, since it is symptomatic of a wider critical acknowledgment in Italy, a recent issue of the literary journal *Allegoria* (Anno XXIII, n. 63, gennaio–giugno 2011) is dedicated to the theme of *Il Modernismo in Italia*. While one certainly welcomes the fact that more worthy Italian scholars are finally willing to question, investigate, and use the term without apparent prejudices, it is still a little disconcerting that futurism's own contribution to this cultural moment is still downplayed (since it is just associated with the avant-garde).

processed,[2] then food and the semantic field around it seem to constitute especially privileged angles from which to approach this period. I argue that the alimentary lens allows to shed new light on the selected texts and that these, in turn, effectively illustrate both the signifying power of the culinary sign, its interconnectedness with aesthetics, and its relevance to emblematizing central concerns of Western modernist sensibility. I shall briefly return later in this introduction to some of the reasons behind my adoption of the term "modernism" in the Italian context and, at the same time, the peculiarly Italian reluctance to use it to define the indigenous literature produced from the beginning of the twentieth century to the end of World War II. Let me begin by providing some further justifications for my topic and addressing at least a few of the interrelated questions that it may raise.

Some Reasons Behind a Choice

First and foremost: why focus on food? What can an examination of the alimentary dimension in the work of these five Italian writers tell us that we do not already know about them and the cultural milieu they belong to? Second, what is the relation between these figures – how and why may be they part of a similar discourse? And, finally, why choose these writers and not others in the same period who made an equal or, arguably, even more substantial impact on modern Italian (and European) literature and culture?

Leaving momentarily aside the fashionable "hotness" of food as a topic of investigation and its centrality in global discourses about the environment, health, and sustainability and focussing merely on its academic, scholarly dimension, one may begin to address these queries by observing

2 Georges Bataille, *Visions of Excess: Selected Writings, 1927–1939*. Edited and with an Introduction by Allan Stoekl (Minneapolis: University of Minnesota Press, 1985), 59.

that virtually every major twentieth century Western intellectual from Freud onward, has reflected on the multiple cultural roles and implications of food and eating. The following brief list, although certainly incomplete, should help make this point. Freud's understanding of food as part of the civilization process, both linked to the pleasure and to the death principles; his realization that the domain of the oral immediately forces us to distinguish what is and what is not a part of one's self and, more in general, his "portrait of humanness that consists [...] of a psychic interdependency [...] that negotiates love and hate, desire and dread, incorporative identity and immemorial loss, nurturing presence and devouring destruction",[3] provide a more or less explicit foundation for many of the scholars who would subsequently address the subject of food.

Roland Barthes, for example, the author of this introduction's epigraph, began writing seriously about wine, milk, steak and chips in *Mythologies* (1957) and then, in "Toward a Psychosociology of Contemporary Food Consumption," developed an entire semiotic theory centered on food's role in culture. In the mid-1960s, came the work of structuralist scholar Claude Lévi-Strauss in *The Raw and the Cooked*, and of social anthropologist Mary Douglas in *Purity and Danger*. Around the same time, literary theorist Mikhail Bakhtin, while discussing the "grotesque body" in *Rabelais and His World*, famously wrote that its interactions with the world are revealed "most fully and concretely" in the act of eating, because the "body transgresses here its own limits, [...] Here man tastes the world, introduces it into his body, makes it part of himself."[4] In the 1980s, philosopher Julia Kristeva, in her *Powers of Horror: An Essay on Abjection*, gave a psychoanalytical twist to some of Bakhtin's insights, addressing the issue of food loathing, the abject, and their relevance for textual interpretation, while Louis Marin, in *Food for Thought* (1986), discussed the communicative power and eminently textual dimension of food, underlying in particular its

3 Susan Skubal, *Word of Mouth. Food and Fiction after Freud* (New York and London: Routledge, 2002), 4.

4 Quoted in Sue Vice, *Introducing Bakhtin* (Manchester and New York: Manchester University Press, 1997), 165.

trans-signifying capacity. And finally, within these same decades, additional widely influential thinkers such as Foucault, Levinas, Derrida, Agamben, and Deleuze and Guattari eventually addressed the trope of alimentation in order to rethink the relationships between the subject and the Other and, in short, to "thematize questions of what is excluded from the subject, what sustains the subject, and what undoes the subject."[5] All these scholars made important contributions to the field of food studies/food scholarship (also known as "gastro-criticism"), implicitly underlying the cultural pervasiveness and interdisciplinary nature of the field and providing the basis, both in terms of theoretical tools and methodological approaches, for future scholarly explorations including, of course, the present one.[6]

While it does not seem necessary to reassess more in depth these historical and relatively well-known contributions to the field in order to reassert its (and their) relevance, it is probably more fruitful to consider a sample of recent explorations (mostly a product of literary critics) which, both in their similarities and divergences, are particularly relevant for my own discussion. This sample, by progressively bringing us closer to the Italian context and the time frame of modernity, will also set the stage for answering my remaining questions.

I start by quoting Italian culinary historian Massimo Montanari's *Food Is Culture* (2006) mostly because of its explicit title, although his work is not literary in its orientation and does not deal with the period I consider. His book, following the approach of Fernand Braudel and the French *Annales* school, immediately helps to explain the vast interdisciplinary interest in the topic and summarizes my own understanding of "what food is". Indeed, Montanari's book points right away to food's relevance as a polysemous instrument for cultural analysis, its historical centrality and significance

5 In: Julie R. Klein. "Nature's Metabolism: On Eating in Derrida, Agamben, and Spinoza" in *Research in Phenomenology*, 33 Koninklijke Brill NV, Leiden, The Netherlands (2003), 187–217 (187).

6 As will become more apparent later on, my study has also greatly benefited from the work of Carolyn Korsmeyer on taste and, in particular, of Mark Anderson, Leslie Heywood and Maud Ellmann, on the links between modernism and eating disorders.

for any culture not only as a mere byproduct "but one that gives shape to the *mentalités* that structure thought and expression" and, as a direct consequence, something that unavoidably affects also "the imagination and the imaginary arts", literature obviously included.[7]

If, paraphrasing Marx's still sensible thoughts on historical materialism, food and eating are fundamental to the life and well-being of a specific sociocultural system, it is likely that the multifaceted artistic production of such a system will eventually manifest some sign of their importance. Not surprisingly, even within the small field of Italian literature and its historical canon, the literal and metaphorical references to food from medieval to contemporary times (i.e., from Dante Alighieri's pre-eminent alimentary text, *Convivio* (1304–7), in which hunger is both corporeal and intellectual, to Igiaba Scego's short story's "*Salsicce*" (2005), where eating is connected to current issues of national identity and Italianness) are so many and varied that it would be absurd to try conceiving a list of them.

Gian Paolo Biasin's *The Flavors of Modernity: Food and the Novel* (1993), however, exemplifies how it both makes scholarly sense and is critically productive not only to address the general, mutually enriching dynamics between food and literature, but also to narrow the analysis of the culinary sign to a specific genre and historical moment: in this instance, the realistic Italian novel and the period of its birth and expansion from the early nineteenth to the late twentieth century. In his work, Biasin makes a distinction between the realistic or mimetic function of food, which metonymically connects the text to the contextual reality outside of it, and the narrative function which refers to the role of catalyst food has in the development of the story action and its cognitive power.[8] My own work

7 Kara K. Keeling and Scott T. Pollard, eds, *Critical Approaches to Food in Children's Literature* (New York and London: Routledge, 2008), 4–5. For a popular audience friendly approach to the topic, see also John Dickie, *Delizia! The Epic History of Italians and Their Food* (London: Hodder & Stoughton, 2007).

8 Gian Paolo Biasin, *I sapori della modernità* (Bologna: Il Mulino, 1991), 16–28. Biasin also mentions the "connotative" function of food, linked to the "narrative" one, which is useful to characterize the characters at the social, psychological, or affective level.

has certainly been inspired by Biasin's model and the distinction mentioned above is still valid and useful for any analysis of the subject.

I share Biasin's valorization of the culinary sign, his multidisciplinary approach, and his intention to avoid a systematic historical investigation and to adopt a single, inflexible critical perspective. But my discussion focuses on a selection of completely different writers and prevalently non-realistic texts; it does not consider the novel exclusively as the hegemonic genre of modernity, and thus focuses on other cultural artifacts belonging to the period. Ultimately, my analysis relies less on the critical tools of stylistics and narratology and draws more, instead, from the insights of both more recent scholarship and philosophers (Nietzsche, Derrida, Perniola) whose thought, although not always directly dealing with the alimentary sphere, resonates productively with the writers and works under consideration.

Besides Biasin's, at least two other recent books should be mentioned in order to better frame my discussion and clarify why studying the aesthetic representations and implications of food in a variety of texts belonging to the period of modernity may be a particularly rewarding, not to mention satisfying, gesture.

Michel Delville's *Food, Poetry, and the Aesthetic of Consumption: Eating the Avant-Garde* (2008) focuses on the crucial role of food – both "materially and conceptually" – in the history of the Western avant-garde and considers how art and food intersect in the work of a variety of international artists belonging to diverse nationalities and artistic disciplines, from Gertrude Stein and Marcel Duchamp to Andy Warhol and Piero Manzoni.[9] Delville, like myself, recognizes modernism's fascination and anxiety with issues of corporeality, its desire, quoting Tim Armstrong, to "intervene in the body; to render it part of modernity by techniques which may be biological, mechanical, or behavioural."[10] He also argues

9 Michel Delville, *Food, Poetry, and the Aesthetic of Consumption: Eating the Avant-Garde* (New York: Routledge, 2008), 5.

10 Tim Armstrong, *Modernism, Technology and the Body* (Cambridge: Cambridge University Press, 1998), 6. Armstrong notes that in the context of modernity "the body is re-energized, re-formed, subject to new modes of production, representation, and commodification" (2).

that to better understand the consumptive aesthetics of the avant-garde one should re-evaluate and focus on the hierarchically lower senses of taste and smell as they are engaged by food, rather than merely on the impact of early twentieth-century technology on the higher ones of sight and hearing.[11] Although I share Delville's interest in addressing the aesthetic side of food as it relates to the body, language, and subjectivity, his book dedicates only a few pages to the most famous alimentary text of Italian futurism, *The Futurist Cookbook* (1932), therefore leaving most of my area of investigation practically untouched.

Finally, Cecilia Novero's *Antidiets of the Avant-Garde: From Futurist Cooking to Eat-Art* (2010) examines how the metaphorical and material presence of food and consumption in the avant-garde at the beginning of the twentieth century then reappears in the neo-avant-garde between the 1960s and 1980s, thus revealing "a temporality of returns that does not preclude aesthetic change".[12] Although from a different methodological angle and choice of texts and authors, her linking the ideas of incorporation (in Walter Benjamin's writings on eating and cooking) and indigestion (in Dada) to a "crisis of communication of the aesthetic experience" seems to complement some of my own reflections (and Delville's too, in his chapter on "Minimalists and Anorexics") on the connection between corporeal self-consumption, modernist language crisis, and aesthetic creation. Expectedly, Novero dedicates her first chapter to the Futurist banquets and the *Futurist Cookbook*, and she provides an accurate historical overview of the period and cultural practice, including insightful observations and close-readings of some futurist recipes. Once again, however, as in Delville's work and differently from mine, her focus is limited to the most obvious futurist text about food and she does not explore other manifestations of the culinary either within futurism or elsewhere in the literature or arts contemporary to it. In her following chapters, moreover, her discussion

11 On the relationship between modernism and the senses, see Sarah Danius, *The Senses of Modernism: Technology, Perception and Aesthetics* (Ithaca: Cornell University Press, 2002).

12 Cecilia Novero, *Antidiets of the Avant-Garde: From Futurist Cooking to Eat-Art* (Minneapolis and London: University of Minnesota Press, 2010), 257.

diverges even further from mine as it shifts towards the neo-avant-gardes, art history, the visual and performance arts.

The books I mention reaffirm not only the relevance of the culinary as an instrument for literary and cultural analysis, but also the logic and the advantage of using it as a key to investigate texts belonging to two of the categories that scholars commonly use to define (also) Italian cultural history on the cusp of the nineteenth and twentieth century: respectively, modernity for Biasin, and the "avant-garde" for Novero.

There are few doubts, however, that as soon as we move away from the limits of the Italian critical landscape, "modernism" seems to be the preferred term to more comprehensively indicate the contradictions and complexity of this period's culture. As Michael Levenson notes, this notion immediately correlates these years – and one should also clarify, be they referring to Italy or elsewhere in Europe – to a generalized "sense of crisis." This crisis could be partially described by the following presences (and implicit absences): the presence of experimental attitudes that wish to break with traditions and conventions; of growing doubts about the links between words and things and the potential of language to process reality; of a new relationship between artists and bourgeois society; of a progressive shift in economies from production to consumption; of a "sentiment of laceration [...] a loss of coherence, weight, and depth of the subject" (Ceserani, in Somigli-Moroni, 54) and – to conclude with a further nod to my main argument – of an acknowledgement of both the potential and limits of the body to "digest" and thus represent a world that is changing at an unprecedented pace.

While I do not wish to provide one more definition of either modernity, the avant-garde, or modernism, I share those critical positions that do not emphasize a definite ideological and aesthetic chasm between these notions but, rather consider them as entwined and correlated experiences which, taken together, help better understand modern life and culture as a whole (in Italy too, notwithstanding its peculiarities).[13] My choice of adopting the all-encompassing modernist label, therefore, reflects and participates in recent scholarly efforts arguing that the literature and culture produced in Italy from the time of its unification in 1861 until the late-1930s does not particularly benefit from being subdivided in its historical partitions (i.e., "symbolism," "decadentism," "crepuscularism," "futurism," "*novecentismo*," etc.).[14] That is, without erasing the importance of local traditions and the particular distinctions between artistic figures and practices, it seems nonetheless more advantageous, from a critical and cultural perspective, to adopt the category of modernism for the Italian case too, since this gesture brings the related discourse more in tune with that of other European literatures.[15]

One additional, practical advantage of this stance convenient for this study is that it "allows us to bring into significant relation experiences which have traditionally been kept separate in Italian criticism" (Somigli-Moroni 12) and therefore binds together artists that, otherwise, would remain isolated under their own separate literary-didactic-critical tag. That is – and I should thus get to address my other two initial questions – my

13 See, for example, Jonathan Crary who affirms that "any effective account of modern culture must confront the ways in which modernism, rather than being a reaction against or transcendence of processes of scientific and economic rationalization, is inseparable from them." Jonathan Crary, *Techniques of the Observer: On Vision and Modernity in the Nineteenth Century* (Cambridge: MIT Press 1992), 85. On a similar position, see also M. Bradbury and J. McFarlane and M. Berman.

14 Paolo Valesio put it this way: "Italian literature has not seemed very interested in the idea of 'selling' itself, as far as its history after the end of the sixteenth century is concerned." (in Somigli and Moroni 2004, x).

15 Luca Somigli and Mario Moroni have already well indicated the historical reasons for the peculiarly Italian hostility to the term modernism. Among these, they individuate "the prevalence [among post-unification intellectuals] of a scientific mentality [that] favored the formation of a naturalistic literature" (15); the impossibility for highly influential artists such as D'Annunzio and Carducci to "conceive literary modernity outside the context of Classicism" (17), and the role that Fascism's totalitarian and repressive regime had in weakening "the ties linking Italian modernism to its European counterparts" (Somigli and Moroni 2004, 13/15/17).

inclusion of figures as profoundly diverse in style, form, and poetics such as Marinetti and Masino, or the early, avant-gardist Palazzeschi and the late Bontempelli or Pirandello (and also the mix of genres, novels, poems, short stories, and plays) may be clarified not only by the fact that each of these artists creatively exploits the signifying power of food, or that they exemplify, each in their own way, some of the intersections between writing, the body, and consumption (or non-consumption), but also that they all are, first and foremost, quintessential modernists writers. This said, one could still question why I selected these specific figures for discussion rather than some other, contemporary Italian author in which both the trope of food and the corporeal dimension could be considered central issues (one may think, for example, at Gadda and his poetic of "macaronic pastiche," or Svevo's inept and failing, "border-line" bodies).[16] My answer is that I wished to focus on texts which, with the exception of those by Pirandello, have long been marginalized and/or barely approached by the Italian critical orthodoxy and to propose an alternative, idiosyncratic modernist "line" (i.e., Marinetti – Pirandello) that could perhaps, provocatively, both establish a dialogue with and run parallel to the canonical one "Svevo-Pirandello" notoriously discussed in a classic essay by Renato Barilli.[17]

With the exception of the younger Masino (born in 1908), the remaining writers I focus on were born around the same time in the third quarter of the nineteenth century, and they were most likely aware of, if not often reciprocally influenced by, each other's artistic production. To give an idea of the synchronicity and interrelatedness of these figures, Palazzeschi and Marinetti were initially fellow futurists; Bontempelli's involvement with the poetics of "return to order" was partially a reaction towards Marinetti's and

16 Interestingly, in a 1928 letter to B. Crémieux, Svevo (who was from Trieste) uses the following alimentary similitude to describe his own feeling of marginality within the Italian literary establishment: "un pezzo d'aglio nella cucina di gente che non ne vogliono sapere." In: Italo Svevo, *Opera omnia I. Epistolario* (Milan: Dall'Oglio 1966), 874.

17 Renato Barilli, *La linea Svevo – Pirandello* (Milano: Mursia, 1972). Simona Cigliana's recent article "Pirandello, futurista 'a suo modo'", in *Italian Culture* XXII (2004), 21–45, further confirms the legitimacy of my suggestion.

futurism's "excesses";[18] Masino was Bontempelli's life companion and both of them were very close to Pirandello; and, finally, Marinetti, Pirandello, and Bontempelli were all among the first members of the Reale Accademia d'Italia (Royal Italian Academy), founded by Mussolini in 1929.

One should note that it is precisely during fascism's cultural policies that this group of four men and one woman lived a substantial part of their artistic lives. Now, the Italian situation during the post WWI years and fascism certainly was, for a minority, a moment of exhilaration and voracious conspicuous consumption. There are no doubts that the regime could support literary careers, facilitate publications and sales and, in short, make intellectuals who were close to it both famous and wealthy. For many others, however (including, for a time, Bontempelli and Masino themselves) this was also a time of suffering and real hunger. These were decades when the combination of economic and alimentary concerns could cause concrete changes at the social and political level and, in turn, also greatly stimulate the most imaginative and artistic minds of the time.

In this scenario, one of the advantages of assigning to the notion of consumption both an economic and an aesthetic/alimentary meaning, that is, of linking it to the physical body and the nutritional flows that (should) regulate its equilibrium, is that it may render it an even more powerful source of crucial information on the consumer/artist and her/his sociopolitical contexts.[19] Within the specific modernist milieu, its exploration implies dealing simultaneously with issues regarding the private and the public, the sociopolitical as well as the individual body, the economies of the self and those of the wider world in which the subject operates. After all, as Mary Douglas wrote, we should be "prepared to see in the body a

18 The expression "return to order" indicates the poetics expressed in the journal *La Ronda* (1919–23), in particular, the desire to go back to the guiding principles of classical art and "the institutionalization of the separation between art and literature and politics" (Somigli and Moroni 2004, 21).

19 In his own reflections on the notion of consumption, Tim Armstrong, quoting Lawrence Birkin, writes that "a de-sublimated appetitive body seems to be the engine of economic modernity" (Armstrong 9). This additional alimentary reference, though indirect, supports further my argument.

symbol of society, and to see the powers and dangers credited to social structure reproduced in small on the human body."[20]

In Italy, the various battles (for grain, milk, rice, grapes) that fascism initiated around the mid-twenties are further proof that the question of alimentation was crucial to the economies of the human and the social bodies and their relationship. In this historical context not only was there "an unbounded mixing of biology with politics [...] individuals were to identify with the nation as cells to an organism [...] there was thought to be no distinction between culture and biology, morality and physiology," but also the emblematic image of a breastfeeding mother could become a "central unifying concept of fascist domestic politics as well as a powerful symbol" of fascism itself.[21] Many among the texts I consider exploit the symbolic richness and metaphorical potential of the alimentary dimension to reflect, illustrate, and comment upon related dynamics and tensions between private and public bodies.

I take Marshall Berman's careful choice of terms literally when, summarizing Karl Marx's view on the opposite forces that will shape the culture of modernism, he lists, among others, "the theme of *insatiable* desires and drives [...] perpetual creation and renewal in every sphere of life; and its radical antithesis, the theme of nihilism, *insatiable* destruction, *the shattering and swallowing up* of life, the heart of darkness, the horror [italics mine]."[22] In other words, I interpret his words as if they suggest that metaphorical images of consumption, in all its connotations, are at the very roots of modernist culture itself.

Although from different perspectives and with varied energy and success, the writers I consider relate and creatively respond – with an intensity

20 Mary Douglas, *Purity and Danger: An Analysis of the Concepts of Pollution and Taboo* (London and New York: Routledge, 1991), 115.

21 Emily Dixon Whitaker, *Measuring Mamma's Milk: Fascism and the Medicalization of Maternity in Italy* (Ann Arbor: University of Michigan Press, 2000), 95–6. The author also notes how "Biological politics aimed to rationalize and unify production and reproduction, fusing biology with economics" (96).

22 Marshall Berman, *All That Is Solid Melts Into Air. The Experience of Modernity* (New York: Penguin Books, 1982), 102.

and depth similar to those many of their European counterparts – to the modern epistemological crisis of subjectivity and representation and to the multiple challenges and threats that industrial modernity (its Italian belatedness notwithstanding), totalitarian regimes, and market capitalism pose to individual and social bodies. Within the limits of an analysis that, ultimately, wishes to remain literary and empirically close to the primary sources, and balanced between the physical and the aesthetic sides, I contend that focusing on the alimentary dimension as these purposely non-homogeneous, gender-diverse, and yet also intertwined writers artistically interpret it, may help illuminate precisely the nature of such common challenges and threats.

Thus, Marinetti – among the first to capitalize on the work of art's heteronomy and consumable character – displays a bold, aggressive, assimilative rhetoric aimed at maintaining "control and possession" of the world, "expresses futurism's evolutionist agenda in alimentary terms"[23] and, contrary to his own declared intentions, constantly recycles and incorporates his own former ideas into the production of allegedly new ones. At the same time, Palazzeschi's references to food in his earlier, avant-garde poems and novels stimulate parallels with notions such as "rejection," "emptiness," "self-effacement" and, implicitly, with the fragility and evanescence of the subject in this cultural-ideological context. A few years later then, a terrifying mythical and destructive image of cannibalism – Saturn devouring his children – will be, paradoxically, still at the core of one of Bontempelli's major novels *Gente nel tempo* and of his project of re-envisioning and "reconstructing" the space and time of modernity precisely through the inspiration provided by ancient myths. The events in this novel, however, seem to indicate that such a project is both contradictory and, literally, physically consuming for those involved in it. Finally, additional representations and variations of hunger and consumption reappear in the remaining works by Bontempelli, Masino, and Pirandello, in which they are investigated also in the light of their relationship with issues of gender and genre.

23 Cinzia Sartini Blum, *The Other Modernism. F.T. Marinetti's Futurist Fiction of Power* (Berkeley and Los Angeles: University of California Press, 1996), 94–5.

I am not the first investigating the overlapping of corporeal and textual practices, or the relationship that eating (and, alternatively, starving) bodies have with language and verbal expression. As we can see from his quote in *Le Plaisir du texte* that I placed at the beginning of this chapter, Barthes made particularly clear the fact that both eating and writing are grounded in the body. Several other scholars have reiterated this point from various perspectives: from Deleuze and Guattari who, when discussing Kafka in *Toward a Minor Literature* (1986), stated that "writing transform[s] words into things capable of competing with food [...] To speak, and above all to write, is to fast." (20) to the more recent work of scholars such as Mark Anderson, Leslie Heywood, and Maud Ellmann, who address the intimate conjunctions between texts and bodies and, in some occasions, the specific intersections between the notion of a modernist "language crisis" and various forms of eating disorders.[24] As a matter of fact, since some of the insights of Anderson and Heywood on such intersections are closely related and will be repeatedly applied, in conjunction with additional theoretical support, as interpretative keys for a few of the texts examined in the following chapters, it seems worth anticipating here their respective, main arguments.

Anderson's article discusses some iconic modernist writers (Hugo von Hofmannstahl, Franz Kafka, Herman Melville) and notes that some of their most representative literary works display a "minimalist" impulse "towards increasingly brief, laconic, self-cancelling forms of utterance" (39). This is a sign of an anguish deriving from the perceived disconnection between reality and the words necessary to interpret it, and equally affects the linguistic and the corporeal dimensions. Such forms of literary expression – he observes – are often "figured in terms of a physical disgust with food, or prolonged fasting despite the absence of external constraint,

24 See Leslie Heywood, *Dedication to Hunger: the Anorexic Aesthetic in Modern Culture* (Berkeley: University of California, 1996), in particular pages 60–110; Mark Anderson, "Anorexia and Modernism, or How I Learned to Diet in All Directions" in "Discourse: Berkeley Journal for Theoretical Studies in Media and Culture", vol. 11, n. 1, 1989, 28–41; Maud Ellmann, *The Hunger Artists. Starving, Writing and Imprisonment* (Cambridge: Harvard University Press, 1993).

[and] finally of a wasting away of the corporeal self" (29). It is thus possible for him to justify a connection between the field of language and the one of food or, as he puts it, between minimalist, modernist writers and "minimalist anorexics" (regardless of questions of gender), based on the presence of a common "extreme concern with eliminating the text or the body" and "strategies of self-denial and self-negation that seek to establish a primal unity uncontaminated by the 'filth' of the other: sexual differentiation, social hierarchy and power relations, temporality and 'history'" (37).

Heywood's book, especially her second chapter, "From Female Disease to Textual Ideal, or What Modernism Got to Do with It", in turn, picks up and develops some of Anderson's essential points on the relationship between literary modernism and the anorexic ideal in her own analyses of texts by Kafka, Eliot, Pound and Williams. At the same time, however, she is mostly interested in gender issues and, therefore, gives a marked feminist twist to her discussion claiming, for instance, that literary modernism has an "ambivalent, even hostile relation to [the] body as feminine and maternal" (63), or that, more specifically, Kafka's artistic creativity as it emerges from his *Letters to Felice* and "A Hunger Artist" stems out of his detachment from an "empirical existence [...] gendered feminine" which needs to be "eliminated as 'fat', the female body supplanted by the male textual body" (71). Although these remarks seem to indicate a tendency to impose, not always convincingly, a subjective, feminist ideology on the materials she is dealing with – her argument could actually be stronger in its absence – she provides some insightful and thought-provoking analyses which, as it will become evident in the following pages, lend themselves to be potentially applied and adapted to comparable textual circumstances. My work thus hopes to add some new and original materials to this ongoing discussion. While I maintain a cross-literary and international critical stance and seek to establish parallels and references beyond the Italian borders, my main focus is on the Italian cultural context and a group of writers and texts that, for the most part, have not previously been approached from this angle.

The chapters that follow map the intersections between the culinary sign and modernist literature in a selection of works by writers who both concretely shape and imaginatively react to the plethora of social, economic, technological, aesthetic, and epistemological shifts that characterize the complex and contradictory experience of Italian modernity.

After observing that Marinetti's interest in food imagery is present from the very beginning rather than being only a late addition in his poetics, as many still claim, chapter one aptly opens with a discussion of his 1905 (pre-futurist) play *Le Roi Bombance*. The analysis of its metaphors of bodily incorporation and consumption allows me to link its gastronomical and sociopolitical sides and to ultimately claim that this text simultaneously belongs to and undermines the utopian genre. In the next two section of this chapter I expand from two different alimentary perspectives the discourse on futurist utopia I introduced in the initial section. First, I confront too the quintessential futurist culinary utopia, *The Futurist Cookbook* (1932), and claim that, on the one hand, the text resonates with Charles Fourier's earlier "gastrosophic" theories and, on the other, is imbued with the scientific rhetoric of hygienism and, in turn, with the discourse of eugenics, a crucial subtext of the nazi-fascist ideology. Second, I address Marinetti's erotic fiction from the late 1910s and 1920s and argue that he attempts to overcome the body's biological limitations by rhetorically linking it to an inorganic, rather than organic entity. Following up on the controversial association between futurism and fascism and futurism's obsession with matter, I end this chapter with an essay on *Patriotismo insetticida. Romanzo d'avventure legislative* [*Insecticidal Patriotism. A Novel of Legislative Adventures*] (1939), a basically unknown allegorical novel Marinetti wrote in the mid-thirties, shortly following fascism's colonial enterprise in Africa. By analyzing the anthropophagic motif in this work and its correspondence with fascist colonialism's metaphorical "eating" of foreign territories, I illustrate how the trope of cannibalism is actually used here to subtly criticize fascist ideology.

Focusing my analysis on his trilogy of "extraordinary" modernist novels, *: riflessi* (1908), *Il codice di Perelà* (1910) and *La piramide* (1912–14), as well as some of his poetry and his posthumously published novel *Interrogatorio della Contessa Maria* (1925), chapter two addresses Aldo Palazzeschi's vanishing and food-averse bodies in the light of his own notion of "lightness" and the "counter-philosophy" of his three avant-garde manifestoes. I argue that the recurring trope of self-effacement not only seems to betray, in line with Huysmans, a view of aesthetic creation as an act of "evaporation" but also reflects a quintessentially modernist undermining of the notion of the subject as the historical locus of Western rationality. I begin by investigating the "anorexic logic" which arguably pervades *: riflessi* and claim that here too, in a pattern similar to the one Heywood individuates in some of Franz Kafka's works, we are facing an instance of transformation by reduction of the self in order to make art possible. Kafka's short story "A Hunger Artist" continues to provide the comparative background for the next section. Here I mostly focus on *Il codice di Perelà* and suggest that both Palazzeschi and Kafka share a common belief in the precariousness of art as a mode of expression, and in the paradoxical dependence of its value and communicative power precisely on the sensation of precariousness and transience it inspires. Along similar lines, in the following part, the critical discussion addresses the meta-literary quality of Palazzeschi's 1926 novel *La piramide* by paying special attention to its many references to bodily incorporation and evacuation. After focusing on several starving and disappearing bodies, the final section of this chapter introduces instead the notion of "voraciousness" and borrows its theoretical framework from gender studies. My revisionary interpretation of *Interrogatorio della Contessa Maria* (1925) claims that the alimentary rhetoric in this novel not only situates femininity at the center of modernity but also, paradoxically, contributes to the repositioning of that same femininity in an anti-modern dimension.

A continuing attention to the relation between issues of food and gender characterizes also chapter three, in which, after Marinetti and Palazzeschi, I introduce another "power-couple" of Italian modernism: Paola Masino and Massimo Bontempelli. In the first section, I comparatively consider two of their works: the short story *Fame* (1933) by Masino, and Bontempelli's almost homonymous three-act theatrical *pièce La Fame* (1934). As their titles suggest, these texts constitute an appropriate and fertile ground on which to reflect on the creative relationship and gender-influenced poetics of two modernist artists who were intimately, and yet differently, dealing with issues of representation, expression and creativity – not to mention with their own ambitions and antagonistic appetite for affirmation and recognition in the Italian literary and publishing landscape of the period.

The last section of the chapter addresses Paola Masino's novel *Nascita e morte della massaia* (1939). By focusing on the widespread food-imagery in the text, on the progressive "extinction of personality" which marks the life of the *massaia*, and on the opposition between notions of productivity and sterility, I suggest that *Nascita* too may (also) be read as an allegory for the difficulties associated with literary creation. Ultimately – I argue – by implicating that the production of a new kind of artistic artifact may only come after an act of self-consumption, the text follows a "food-averse logic" very similar to that characterizing some of the other modernist texts previously analyzed (i.e., Kafka's and Palazzeschi's).

Brought closer not only, as previously mentioned, by biographical circumstances and their common friendship with Paola Masino but, especially in the decades under consideration, also by the shared belief in the redemptive power of myth in artistic creation as a way to mitigate and compensate the more destructive and dehumanizing effects of modernity, Bontempelli and Pirandello nonetheless present us with different aesthetic solutions to the crisis of subjective wholeness. Chapter four investigates what such solutions may be by taking as indicative samples a novel by the former and a long novella by the latter in which the alimentary dimension plays an important role. Thus, I first interpret Bontempelli's *Gente nel tempo* (1937) by expanding upon the implications of the modernized classical Greek myth of Kronos. I argue that this novel, while pointing to modern manifestations of the destructive, devouring pattern of time known as Kronos, hints also at a way in which an alternative, more human time (Kairos) might be salvaged and re-experienced. I contend that Bontempelli's desired "recovery of the individual" and his related aesthetic project of escape from the negativity of History derives from an alternative *kairotic* experience (such as the one of "play") in which time is characterized by its qualitative rather than its quantitative nature and artistic creativity reacquires a pre-eminent, active role.

Finally, I address Pirandello's short story *La balia* (in *Il silenzio*, 1923) that deals with the complex sociopolitical and economic dynamics taking shape at the beginning of the twentieth century both in Italy and in the modern Western world, by providing a controversial representation of maternity. I claim that *La balia* addresses the inevitable, all-consuming rise of a proto-capitalist economy and its devastating effects on the progressively commodified, objectified bodies of the weakest human beings. At the same time, Pirandello provides a critical commentary on a practice – that of breast-feeding – which was among the primary goals and a powerful symbol of contemporary fascist biological politics, and simultaneously addresses the criminal, cannibalistic exploitation that the Italian North exerted towards its own South.

My conclusion reiterates the idea that a "gastro-critical" approach may be conducive to original interpretations and a deeper understanding not only of the texts and authors I consider, but also of other, more or less canonical works belonging to Italian modernist literature. It also considers some of the reasons why this kind of approach may contribute to revitalize the larger field of literary studies as a whole.

CHAPTER ONE

A Futurist Digest: Filippo Tommaso Marinetti's Alimentary Rhetoric

1.1 *Le Roi Bombance*: Between a Dyspeptic Dystopia and an Embryonic Utopia

> La Grande Dypepsie éclatera tout à coup, coincidant avec le refroidissement de la terre stérile![1]
>
> — F.T. MARINETTI, *Le Roi Bombance*

Even within F.T. Marinetti's French, pre-futurist and thus already less well-known literary production, his *tragédie satirique* (or *tragédie hilare*) *Le Roi Bombance*,[2] published by the editions of *Mercure de France* in 1905 and subsequently staged in 1909 at the *Théâtre de l'Oeuvre* in Paris, stands out as a text still very much at the margins of both editorial and critical

1 "Dyspepsia" "Difficulty or derangement of digestion; indigestion; applied towards forms of disorder of the digestive organs, esp. the stomach, usually involving weakness, loss of appetite, and depression of spirit" (*Oxford English Dictionary*). Here I use the term to indicate in particular the deranged, anomalous digestive situation that affects the characters in *Le Roi Bombance*, one in which starvation, engorgement, cannibalism and vomiting represent a cyclic norm.

2 As Pasquale Jannini notes "Tra il 1894 e il 1912 F.T. Marinetti si volle e fu scrittore francese" in *Scritti francesi* (Milan: Mondadori, 1983), 7. The Italian translation by Decio Cinti, *Re Baldoria*, appeared in 1909. The editions of the play used here are: *Le Roi Bombance* (Paris: Société du Mercure de France, 1905) and *Re Baldoria* (Milano: Fratelli Treves, 1920).

attention. Two simple considerations confirm that Marinetti's social-gastronomical play has generally failed to find an audience in both the publishing and scholarly fields.[3] As of today, the only available editions of the text are either the rare original 1905 French edition, or the even more elusive 1909 (and 1920) Italian translation.[4] The second consideration is that only a handful of full-length scholarly articles have dealt with *Le Roi Bombance* in recent times.[5] Although taking different methodological perspectives, these essays invariably discuss Marinetti's work by observing its satirical, socio-political message delivered through the culinary allegory and by underlining in particular the inter-textual relationship it establishes with one of Alfred Jarry's most popular texts, *Ubu roi* (1896). A quick glance at the plot of *Le Roi Bombance* may illuminate both the pervasiveness of this allegory and, in general terms, the nature of the frequent comparison with Jarry's text:

> The cook (Ripaille) of the imaginary kingdom of Bourdes, who was able to prepare delicious food for his king, has died and his death has initiated a political crisis. The people protest, and the three royal cooks, in order to calm the situation down, promise to organize a banquet. However, their real goal is just to barricade themselves into the royal kitchens and stuff themselves with food. The "starving" crowd discovers that it has been fooled and, increasingly mad, eats the king (Roi Bombance), his cooks, his servants and also the court poet (L'Idiot), the one who, telling his stories, had vainly attempted to direct the appetites of the people towards more idealistic

3 Brunella Eruli notes that "da un punto di vista strettamente teatrale *Roi Bombance* è quanto di meno innovativo e rappresentabile si possa immaginare. [...] teatro da leggere più che da rappresentare" in: "Da un re all'altro: Jarry e Marinetti", *Franco-Italica: Serie Storico-Letteraria* 1 (1992), 145–60, (152).

4 The text is now available in F.T. Marinetti, *Teatro*. A cura di Jeffrey T. Schnapp (Milano: Mondadori, 2004).

5 Here they are: Franco Vazzoler "Marinetti et Jarry," in Marcadé, J.C., ed., *Présence de F.T. Marinetti: actes du colloque international tenu à l'UNESCO* (Lausanne: L'Age d'Homme, 1982), 134–9; Brunella Eruli, "Preistoria francese del futurismo," *Rivista di letterature moderne e comparate* 23, 4 (1970), 245–90; Pierre Gobin, "Goinfrerie et pouvoir: D'*Ubu roi (1896)* au *Roi Bombance (1905/1909)*," *Recherches Semiotiques* 14: 1–2 (1994), 251–67. Cinzia Sartini-Blum, moreover, dedicates few, but enlightening pages (12–14; 93–4) to Marinetti's play in her *The Other Modernism* (1996).

> objectives. In the end, the characters of Sainte Pourriture (Saint Putrefaction) the eternal force of transformation, accompanied by the vampire Ptiokaroum, resuscitates the characters, who are consequently "vomited out" by the same people who had previously ingurgitated them (qtd. in Eruli 1970, 261).

Antipodal observations easily arise after reading this play. It is hard to deny that we are dealing with an almost illegible (and "un-stageable") text, characterized by "lunghissime tirate, linguaggio enfatico e molto tradizionale, azioni macchinose" (Eruli 1992, 152), which rightly deserves to have slipped into oblivion. But one has the impression that the wealth of implications simultaneously elicited and hidden in its verbal and visual overflow may have not been explored in their totality.

Scholars, for example, have not much discussed the suggestions of the play's fairy-tale like quality. The plot, after all, also tells the story of how one of the "king's two bodies" (the royal-historical one) is degraded into the other (the natural-pathetic one) until it overlaps with the classic, fairy-tale like figure of an ogre (Bombance is described as having a "vaste nez bourgeonnant [...] une houppe de cheveux blonds enfarinés [...] la bedaine qui surplombe les cuisses").[6] This visual coincidence may anticipate and make evident that eating and incorporation shall be the true protagonists of the play and that, consequently, the borders of subjectivity are potentially at stake. The relationship between the corporeal and the temporal dimension (or the "body" and "history"), not to mention the one between the "beast" and the "god", already constitute central preoccupations for Marinetti (and the dynamics between Mafarka and Gazurmah in *Mafarka le futuriste* (1909) well examplifies the latter).[7]

The metamorphic topology of a *pièce* in which the light-filled, edible, and "Cockaignesque" castle of the king ("Château Bombance [...] pareil

6 Marin, Louis, *Food for Thought* (Baltimore and London: The Johns Hopkins University Press, 1989), 210.

7 Appropriately, "the 'ogre', by definition, reigns supreme in the sphere of eating and incorporation. He is nothing short of the absolute, a kind of super-cannibal [...] In the body of the ogre, supernature and infraculture are conjoined and form cannibalism: the ogre is a god who behaves like a beast" (Marin 155–6).

[…] une colossale tourte étincelante, crénelée de sucreries roses [avec] colonnades de chocolat […] terrasses beurrées d'or" [9]) is eventually turned into a dark background setting for a cannibalistic orgiasm ("les ténèbres fumeuses," [191]) seems to have been similarly neglected. Even a superficial glance at the *loci* of the text suggests that an edenic situation of abundance and communion with nature can only be an ephemeral dream, and that the discourse on utopia and/or its subsequent negation may be particularly relevant to any interpretive effort.

Finally, if among the many signs of the modernist discourse on the dynamics and extension of the self one can count "a rejection of nineteenth-century historical and social discourse," and "textual events [which] are repeatedly figured in terms of […] prolonged fasting […] of a wasting away of the corporeal self" (Anderson, 29), then the remarkable role that *Le Roi Bombance* may play in this discourse still needs to be adequately evaluated. Let's not forget, in fact, that we are dealing with a "social-gastronomical" text in which history is reduced to a sort of dyspepsic digestive cycle, progress itself is negated, the king's historical-collective body is effaced in order to leave on the scene only the physicality of his natural body, and, last but not least, the character of the poet kills himself with a sword called "La Succulente."[8]

This said, and once I have (hopefully) managed to whet the reader's appetite for Marinetti's play and the width of its potential ramifications, the focus of this section shall necessarily be narrower. As we encounter them, I shall continue to point out those qualities of *Le Roi Bombance* (in its function of literary text rather than theatrical performance) which contribute to make it not only an exceptional (in the sense of "out of the norm") example of twentieth-century literature, but a representative work of literary modernism and some of its aporias. But given the importance of *Ubu Roi* in particular, and of Jarry's artistic production in general to

8 Paradoxically, given the presence of several orgiastic moments in the play, the dynamics here seem antipodal to those of the "orgiasm," where "what is at issue is the collective body which prevails over the body proper." See Michel Maffesoli, *The Shadow of Dionysus: a Contribution to the Sociology of the Orgy* (Albany: State University of New York Press, 1993), 5.

the composition of *Le Roi Bombance* and the cultural and aesthetic formation of the inventor of futurism,[9] I would like to suggest that Jarry[10] may not exhaust the French genealogical ingredients which contributed to the preparation of Marinetti's allegorical cauldron (and I am not merely thinking of a relatively obvious Rabelaisian influence).

One of the first things that any reader may note when comparing the original French version to the successive Italian translation of the play, is not only that the "colorata espressività della lingua [francese] marinettiana" has left its place to the "livello stilistico enfatico e monocorde" of Decio Cinti's version (Eruli 1992, 156), but also – at a much more immediate level – that the dedications that precede the Italian and the French text are completely different.[11] While the Italian translation (approved by Marinetti) chooses to dedicate *Re Baldoria* "Ai Grandi Cuochi della Felicità Universale. Filippo Turati, Enrico Ferri, Arturo Labriola," Marinetti's original *Le Roi Bombance* is dedicated "*A mon cher mâitre et ami. Paul Adam.*" At least one of the roles of a dedication, besides its standard eulogistic and encomiastic function, is to provide a clue to the way one should read and interpret what comes afterwards (in the primary text). It also seems evident that the interpretive directions imparted by each of the pre-texts – while certainly not in blatant contradiction with each other – appear nonetheless to diverge. With this consideration, and at the risk of stating the obvious, I mean that these directions evoke and stress different issues among those which, later on, the primary text (respectively, the French and the Italian) shall equally address and develop. I am not implying that one should expect essentially

9 It may be superfluous to observe that Jarry is only one among a squadron of French writers and intellectuals who had a substantial influence on F.T. Marinetti's aesthetic and artistic development. The best essay exploring the subject is still Eruli's "Preistoria francese del futurismo."

10 And, with him, Max Jacob: among the sources of *Roi Bombance*, in fact, Eruli includes Jarry's *Ubu Roi* but also the similarly "jarryesque" *Histoire du roi Kaboul ler et du marmiton Gauwain* (1904) by Max Jacob (see Eruli 1992, 154).

11 After Gerard Genette's studies on the "relation between the text and its foretexts" in *Palimpsests. Literature in the Second Degree* (Lincoln and London: University of Nebraska Press, 1997), 399, the semiotic importance of what surrounds the primary text does not need to be stressed any further.

divergent messages from the French original and its Italian translation or, even more daringly, that the basic identity of one version differs substantially from the other. More specifically, I am suggesting that after reading each dedication, the reader's "horizon of expectation" of each primary text may be quite different. In more concrete terms: if one sees that a book is dedicated to Italian political figures, one may expect and look for something specific in its pages. If, on the other hand, a volume is dedicated to a French symbolist active in Paris at the *fin-de-siécle* ("le plus grand méconnu de la génération symboliste?"),[12] interested in the *cabale*, and whose name happens also to be associated with a peculiar utopian *roman*, some other line of interpretation may begin to emerge.

Differing expectations, in turn, may change the balance or re-distribute the emphasis of one's overall interpretation of the text(s), so that what seemed marginal and tangential in one circumstance, may come to the forefront, and vice-versa. Concentrating on the diverse hermeneutical implications of each dedicatory note may be a more rewarding operation if the final objective is not just to highlight the different expectations they suggest to the reader, but, rather, to show how such expectations stem out of a unique artistic vision and together, further enrich and problematize Marinetti's *pièce*.

The appropriateness of the Italian dedication as an aid that situates the audience in a familiar frame of reference, simultaneously politicizes the text, and reveals the text's underlying socio-political bone ("a satirical caricature of socialist ideologies of materialistic progress" (Sartini Blum 11)) is hard to miss and, as such, it did not escape the attention of the scholars who read the text accordingly. The invocation of the three Italian socialist leaders in connection with the culinary reference (the "Grandi Cuochi della Felicità Universale") immediately evokes Marx's fundamentally (anti-utopian) statement when asked to portray the future Communist society:

12 In J. Ann Duncan, *Les romans de Paul Adam: Du Symbolisme littéraire au symbolysme cabalistique* (Berne, Francfort/M, Las Vegas: Peter Lang, 1977), 1.

"I do not write cook-books for the kitchens of the future"[13] and, along with it, the Marxist concept of historical materialism, whose basic postulate – in extremely reductive terms – is that the existence of matter does not depend on a supernatural entity, and that man (himself intended as "evolved matter") needs to feed himself before he can think. It is, in other words, the production and the reproduction of material life that allows man to experience intellectual and social progress. As Marx wrote:

> We must begin by stating the first premise of all human existence and, therefore, of all history, the premise, namely, that men must be in a position to live in order to be able to "make history." But life involves before everything else eating and drinking, housing, clothing and various other things. The first historical act is thus the production of the means to satisfy these needs, the production of material life itself.[14]

Not surprisingly, in the specific Italian context, Antonio Labriola reiterated the same principle in his "Dilucidazione preliminare":

> [...] gli uomini, non per elezione ma perché non potrebbero altrimenti, soddisfano prima certi bisogni elementari, e poi da questi ne sviluppano degli altri, raffinandosi [...] il materialismo della interpretazione storica non è se non il tentativo di rifare nella mente, con metodo, la genesi e la complicazione del vivere umano sviluppatosi attraverso i secoli.[15]

Social ideas and progress cannot be produced unless there is economic "peace of mind," which, in turn, brings an alimentary surplus. By allegorically depicting a situation in which such a surplus is lacking and showing the tragicomic consequences of this absence, *Re Baldoria* manages to corroborate (*e-negativo*) the Marxist-socialist idea that the satisfaction

13 Qtd. in Krishan Kumar, "Utopia and Anti-Utopia in the Twentieth Century", in R. Schaer, G. Claeys, L. Tower Sargent, eds, *Utopia* (New York, Oxford: Oxford University Press, 2000), 251.

14 Karl Marx, *The German Ideology* (Amherst, NY: Prometheus Books, 1998), 47.

15 Labriola's essay constitutes a section of his *Del materialismo storico* (Bari: Laterza, 1971), 63/116. As Jannini (and others) noted (although consistently failing to provide a complete bibliographical reference), Labriola wrote a very positive review of Marinetti's play in the Socialist newspaper *Avanti!* (Jannini 26).

of elementary needs (such as eating) is a necessary element for social and historical progress and that, vice-versa, its unfulfillment can only bring a deadly cycle of repetition of the *status quo* (as the end of the play suggests, with the final re-vomiting of the previously ingested characters and the apparent re-establishment of the initial situation). In this sense, one can say that the play ultimately deals with the motif of the uselessness of any human effort to change things[16] and that Labriola's further consideration in his "Dilucidazione": "di storia come vero processo non ce n'è [...] la storia è una canzone noiosa" (116), could indeed summarize its innermost message.

That the content of such a message, in its essential value, is not subverted, or radically transformed even after one is welcomed into Marinetti's French text by the dedication to Paul Adam (1862–1920) is, for clarity's sake, a crucial premise to the rest of my observations.

One of the things that the French dedication accomplishes differently than its Italian counterpart (together with the introduction of Paul Adam's figure into the picture), is the following. For most readers – I would argue – the dedication does not give away in such a crystal-clear fashion (in comparison to the blatancy of the Italian one) the specific socio-political, allegorical nature of the text it introduces. More in particular, by setting *Le Roi Bombance* in an implicit dialogue with Paul Adam's work (who, in fact, is defined first as a *mâitre* and then as an *ami*) the dedication seems to simply suggest first Marinetti's general familiarity with Adam's *oeuvre* and, second, his creative and artistic debt to the genre of literature which Adam was most famous for, that is: *symbolisme*.[17] The more specific and important

16 See (Eruli 1970, 261).

17 "Trente trois romans, sans compter les trente quatre volumes d'essais ou de contes, les trois pièces de theater ..." (Duncan 3). Marinetti briefly writes about the friendship that linked him to Adam both in his *Una sensibilità italiana nata in Egitto* and in *La grande Milano tradizionale e futurista* (Milano: Mondadori, 1969). In the latter we learn that: "Il mio ingegno ansioso di nuove forme letterarie era carezzato dalla affettuosa amicizia di Paul Adam e con piacere lo visitavo nella sua villa un pomeriggio torrido d'agosto [...] Mi credo solo cosicché sofficemente entrando l'illustre padrone di casa agile ma cubico a barba biondastra di faccia meridionale ottimista e forte sguardo fiero e dolce mi parla del suo libro *Trust* e di un probabile più futurista ...." (223–4). Even though *Le Trust* (1910) is Adam's only novel which he directly

realization that Adam's political vision and social ideas may have actually played a role in shaping Marinetti's play may not be so evident at first sight.

It is only once the real subject-matter of *Le Roi Bombance* as a whole has been grasped that one begins to realize that the play may partially belong to and, at the same time, subvert that genre of allegorical utopian fiction which was experiencing a renaissance in the last decade of the twentieth century in France (paradoxically, thanks to several socialist authors) and to which Adam (himself a "socialiste" although "un peu à sa manière" [Duncan 19]) had also contributed with a curious 1897 *roman*, *Lettres de Malaisie*.[18]

This epistolary novel describes an ideal society founded on an island in the East Indies by a group of French settlers who "fuyant l'Europe par haine de l'injustice et de la guerre sociale" (56) and who are said to be disciples of earlier utopian thinkers such as Fourier, Saint-Simon, Cabet, and Proudhon. The colony is therefore fed and sustained precisely by those progressive myths and socialist ideals of "universal happiness" that evidently constitute the satirical target of Marinetti's play. Adam's work presents us with an ostentatiously comfortable society characterized by order, harmony, and justice, which has managed to escape from material needs thanks to the advancement of science and technology and the development of new ways to increase the production of industry and agriculture. Generally following the stereotypical paradigm of a classic utopian narration, the traveler, a nameless "diplomate espagnol" who is helped to orient himself in this foreign and exotic land by his guides (singularly for the genre, two charming women named Pythie and Thea), illustrates the principal local

---

mentions – interestingly, one whose "leitmotif [...] est celle [...] qu' 'il faut tuer pour créer'" (Duncan 180) – and *Lettres de Malaisie* does not officially enter the picture, one cannot but already wonder if such an epistolary utopian novel could not, in fact, perfectly fulfill both Marinetti's craving for a "nuov[a] form[a] letterari[a]" as well as fit the description of the mysterious [book] "probabile più futurista" which he is elliptically hinting at.

18 Paul Adam, *Lettres de Malaisie* (Paris: G. Crès, 1922). Within Adam's *oeuvre*, his only other literary experience comparable to his *Lettres de Malaisie* was the novel *Coeurs nouveaux*, which was published one year earlier (1896). See Minerva, Nadia, "Une utopie inquiète: Les *Lettres de Malaisie* de Paul Adam (1898)," *L'Esprit Créateur* 34, n. 4 (Winter 1994), 77.

features and habits in nine long letters to the narrator. Through this narrative device the reader is instructed on a wide range of arguments: from the presence of technological marvels or the enviable civic virtues and the political and social organization of the indigenous population, to its religion, its sexual behavior, its flawless justice and education systems, the structure of its army and, last but not least, its eating habits.[19]

As far as structure, form, and content are concerned, we are dealing with a text which could not be more radically distant from Marinetti's. In particular, if *Lettres de Malaisie* aims to illustrate an ideal society which, as any utopian project, attempts to "break with bestial origins," *Le Roi Bombance*, as the dramatic events in the "règne des Bourdes" show, satirically depicts an opposite "fictional reality," one which belongs to a landscape of corporeal degradation and animalistic abjection and, thus, could easily fit the definition of a counter-utopia or dystopia.[20]

At this point, then, one wonders why and in what sense Marinetti still cared to mention Adam as his teacher (*maître*) when, in practice, he apparently parodizes and radically subverts one of Adam's most important teachings and beliefs. When, in other words, he actually chooses in his own play to depict a pessimistic, dystopian reality that is a "satirical caricature" of – and in direct contrast with – not only socialist ideologies in general but Adam's own socialist utopian vision (a vision whose positive representation seems to be the main objective of his *Lettres de Malaisie*).

A closer look at the nature of Adam's novel and at some themes dear to the nineteenth century utopian imagination, themes it coincidentally shares

19 See (Minerva 1994, 80–1). In *Lettres de Malaise*, the references to the culinary preferences and the alimentary organization of the population (see, for example, the sections which describes "les Abattoirs" (156–7) or the one about "les fabriques culinaires" (164)) may establish a first, though still tenuous, bridge with the gastronomical allegory in *Le Roi Bombance*.

20 As Frédéric Rouvillois also noted, "Utopia's hygienic compulsion, its aversion to odors and filth and, by extension, to all recalcitrant forms of nature [...] grows out of a similar tendency [the "break with bestial origins"]. The new self merits a leveled, smooth and transparent environment that fully displays the demiurgic powers of its designer" in Schaer, "Utopia and Totalitarianism", 319. The difference with the Rabelaisian corporeal landscape in *Le Roi Bombance* could not be greater.

with *Le Roi Bombance* (individual vs. mass happiness, anarchy, freedom), may provide a tentative answer to the question asked above. This scrutiny may reveal that what Marinetti-author-of-*Le Roi Bombance* has derived from Adam (and, arguably, directly or indirectly also from his *Lettres de Malaisie*) is, among other things, a lasting interest in the utopian model (as his futurist ideal shall soon confirm) but also a simultaneous *mise en question* of the utopian discourse itself and its necessarily optimistic outlook. If the utopia in *Lettres de Malaisie* is at the same time "prise au sérieux et tournée en ridicule" (Minerva 1994, 83) and, therefore, Adam's novel presents also the traits of a counter-utopia, then one consequence of such a situation is that the gap that separates it from *Le Roi Bombance* becomes narrower, as both works share a similar dystopian vision. Marinetti's reference to Adam as his *mâitre*, by extension, makes more sense if the latter's perspective is somehow shared rather than thoroughly questioned.

The first clue that a dystopian current flows under the pages of Adam's novel is the ambiguity of its title, which subtly plays between the meanings of *Malaisie*, that is, the geographical name of the utopian, exotic country (Malaya), and "malaise" which alludes to a disquieting illness, a discomfort, a feeling and situation of general uneasiness. If, therefore, we are distant from the physical, dyspepsic cycle of gastric "malaise" which allegorically torments the characters in *Le Roi Bombance*, Adam's account is nonetheless also marred, even before its beginning, by the ghostly presence of an unspecified sickness. In addition, as the epistolary description progresses, we realize that the colony's harmony has been achieved at the price of a widespread and oppressive state control. The asocial are sterilized; artists are censored; criminals are enlisted into the Army; and sexuality – one of the most powerful threats to the collective social order – is neutralized through institutionalized public orgies intended to achieve a "communisme de sensations érotiques" aimed at destroying "le désir de proprieté sur l'amante ou sur l'amant" (76) and at extinguishing single individualities and subjectivities. The sort of enantiodromic rhetoric[21] which informs Adam's

21 "Enantiodromia" (clashing together), "is the being torn asunder into pairs of opposites […] the law of enantiodromia, according to which at some time everything

novel has been convincingly analyzed by Nadia Minerva, who demonstrates that, indeed, the modernity of *Lettres de Malaisie* consists especially in the fact that the Malayan civilization is presented as a utopian model, as the grotesque deformation of such a model and, finally, as the reversal of the latter into a true dystopia.[22] The crux of her argument is that Adam creates an "effet utopie" in his work by inter-textually juxtaposing different and often dissonant "imaginaires utopiques" which, however, create a parodical contradiction and, more frequently than not, provoke the distancing doubt or sarcastic response of the narrator. The latter's ambiguous position between appreciation and disapproval of the colony's policies emerges – just to give an example – even in tangential considerations such as the following, in which one witnesses the "malaise de Malaise" concretely coming to the surface: "En vain je demandai la permission d'emporter ma boite à cigars. Mes gardiennes declarerent que l'alcool et le tabac n'avaient point droit de cite dans le pays. Je ressens un malaise de cette privation" (31). As the passage shows, the colony is based on a series of abolitions that often deeply disturb the narrator: from small ones such as forbidding drinking and smoking, to crucial ones such as "la famille, [le] capital, [...] la concurrence, [...] l'amour," up to "la liberté" (26). The result of these constant prohibitions is a negation of the concept of individual freedom for the sake of the collective good and, as such, it is a situation that both the fictional Spanish diplomat and his real creator, the quintessentially still "anarchiste Adam" (Minerva 1994, 85), cannot ultimately cope with.

The skeptical and doubtful attitude of the narrator has the effect of desecrating and disclaiming the "naiveté, l'enthousiasme puéril, l'optimisme trop facile des utopias 'a la mode'" (Minerva 1994, 83). And it is not hard

meets with its opposite" (*OED*). In the present context, therefore, such a term seems useful to synthetically describe a situation where a utopia coexists and meets with a dystopia.

22 See (Minerva 1994) and also her "Il pudore della felicità: luoghi comuni e parole tabù nell'utopia francese tra Otto e Novecento", in Biondi, C., Imbroscio, C. Latil, M.J., Minerva, N., Pellandra, C., Sfragaro, A., Souberyan, B., Vecchi, P., eds, *La quête du bonheur et l'expression de la doleur dans la littérature et la pensée francaises* (Geneva: Droz, 1995), 181–8.

to see that the socialist utopia occupies a privileged place among those fashionable utopias that come under scrutiny within this sort of textual dynamics. Thus, while Adam does illustrate in his novel some of the positive aspects of this kind of utopia, he also manages to convey its excesses and deformations; its potential unexpected and undesirable aspects; its shortcomings and the dangers involved in the practical realization and the search for "universal happiness" (Minerva 1994, 85):

> J'y demontre burlesquenment la victoire tragique et fatale de l'individualisme idéaliste sur la masse brutale. Bref, je conclus avec la faillite du socialisme, la gloire de l'anarchie et la complete ridiculisation des bonimenteurs, reformists et autres "marmitons" du Bonheur universel.[23]

"The victory of the individual over the mass"; "the failure of socialism"; "the glory of anarchy"; the fundamental critique of those who still "sell" the dream of universal happiness: after our brief glance at Adam's novel, these words could ideally belong to the French writer as he describes the themes that his *Lettres de Malaisie* address under their utopian *façade*. Well, as a matter of fact, they do not, as the declamatory tone, the presence of linguistic clues such as the adverb "burlesquement," or the noun "marmitons" reveal. Their author is, expectedly, F.T. Marinetti, who gives here, in a 1905 letter to Giovanni Pascoli, his own interpretation of the meaning of *Le Roi Bombance*. A closer look at the first point he raises, "la victoire tragique et fatale de l'individualisme idéaliste sur la masse brutale," which deals with the crucial question (both in *Le Roi Bombance* and *Lettres de Malaisie*) of the fate of the subject and of individuality in a utopian (or dystopian) context, may allow us to draw some conclusions on the relationship that may link these works.

In the play, this "victory" is achieved as the character of "l'Idiot," the poet who has been trying to orient the mass towards more idealistic goals, kills himself with a self-inflicted sword-blow to the head, after realizing that "liberté," as well as "justice" and "égalité" cannot exist in a static and infinitely repetitive historical situation such as the one in which he finds

23 Qtd. in (Eruli 1992, 153).

himself: "La Liberté? ... Cela ne se mange pas [...] Elle est insaisissable [...] c'est notre essor à tire-d'ailes vers un azur toujours plus large et plus abreuvant!" (250–2). That his death is actually a victory can be probably better understood especially if one recurs to the notion of "modernist literary anorexia." What this notion – arguably another one among those "paradoxes of literary modernity" studied by Guglielmi[24] – entails is that "self-consumption has cleared the way for the production of 'purified art'" or that "the trope of spiritual rebirth [...] is made possible by getting rid of the dependent, the finite" (Heywood 76/81).

As the poet gets rid of his own self, he also eliminates those elements that prevent the fulfillment of his idealistic vision. By this self-effacing gesture, he paradoxically establishes (in the long run) the new premises for the realization of the "ideal," be it an aesthetic one or, in more concrete terms, even a (utopian) socio-political one.[25] As a matter of fact, the final, active, self-negating act performed by the poet seems more efficient, as far as the potential achievement of its objective is concerned – (i.e., the interruption of the "tiresome song" which is History and the subsequent pursue of an ideal) – than the merely rhetorical consideration made by the vampire Ptiokaroum (perhaps, in this circumscribed sense, a comparable character to "l'Idiot") who, equally tired of the monotony of the situation, anticipates a progressive, proto-futurist attitude as he states that "Le devenir, voilà la seule religion! ... Quand vous regretterez quelque chose ... c'est déja un germe de mort que vous portez en vous! ..." (263).

Eruli noted that an "afflato nietzschiano" pervades the final part of the play (Eruli 1992, 157). As Nietzsche reminds us, the cult of history that afflicts modern man is a paralyzing one, as it prevents any active and creative disposition in the present as well as in the future. When the poet opts

24 See Guido Guglielmi, "The Paradoxes of Literary Modernity," *The Yale Journal of Criticism*, 5, n. 1, 1991, 217–33.

25 Eruli's observation is timely: "Il rapporto poeta-mondo prevede [...] o la distruzione del mondo o la distruzione del poeta, e quest'ultima sotto la forma di un volontario annullamento [...] o di riduzione a strumento di una realtà cosmica più vasta e misteriosa cioè l' 'esprit poétique'. L' 'esprit poétique' è la sola facoltà capace di comprendere le leggi dell'analogia universale e di arrivare alla verità" (Eruli 1970, 270).

to allegorically disappear from the text, he finds both a way to undermine the notion of a subject who is the "fondamento della ragione occidentale,"[26] and to escape from the heavy "cage" and the "malattia" of history.[27] His self-effacing mode becomes thus a paradoxical solution with which to end history and, at the same time, free oneself from its sickly and stifling influence. As such, his gesture can be interpreted as not simply nihilistic, since his sacrifice leaves the space empty and ready for the future possibility of an absolute beginning, a new point of origin and/or a re-birth – something which on a different level may be conducive to the realization of any kind of utopian ideal (or a new, better history, so to speak, as temporary as it may be).

Returning to the gastronomical allegory in *Le Roi Bombance*: it is only once the sickness caused by history is defeated via a fatal act of self-annihilation; once the dyspepsic cycle characterized by the cannibalistic engorgement and the subsequent vomiting which contributes to lock the characters into the dystopian *status quo* is drastically (even though temporarily) interrupted, and an opposite, specifically anorexic refusal to eat takes its place, that the illusion of a reconciliation between the world of history and the one of absolute values – or, otherwise, the utopian dream – may have a chance to eventually thrive. It is thus significant that, at the end of the play, the poet commits suicide after saying that freedom is not edible: "La Liberté? ... Cela ne se mange pas!"; and that Ptiokaroum does not want to eat his brain, saying that he has already eaten too many Bourdes: "j'ai fait une indigestion de Bourdes ... Je suis ... fatigue" (267).

Can we suggest, at this point, that an imaginary, indicative proportion which may catch the essence of *Le Roi Bombance* could be one that joins "dyspepsia" and "dystopia" on one side and "anorexia" and "utopia" on the other? Perhaps we can, if we consider that anorexia is in itself another form of sickness but also, in these circumstances, a way to refer to the "modernist

26 See Mario Moroni, *La presenza complessa. Identità e soggettività nelle poetiche del Novecento* (Ravenna: Longo, 1998), 21.

27 On this question, see Guido Guglielmi, *La parola del testo: letteratura come storia* (Bologna: Il Mulino, 1993), 158.

logic" that partially informs Marinetti's text and may paradoxically save it from its otherwise cyclic doom of dyspepsic stasis. And, second, if we realize that such a logic, which ultimately implies that it is only from an absence that a presence may eventually be produced, seems eerily reminiscent of the one that informs any utopian imagination. The power of utopia, in fact, only rests on the absent intangibility of desire and hope and as soon one tries to materially make those a presence (as Adam's work perfectly seems to demonstrate), "elle se suicide: le desir se tait, l'espoir s'évanouit" (Minerva 1994, 87), and dystopia steps in to fill the void.

The emblematic quotation from Colossians mentioned by Frèderic Rouvillois in his essay on "Utopia and Totalitarianism": "You have stripped off the old self ... and have clothed yourself with the new self" may further corroborate the suggested link among the notions of anorexia and utopia. As he notes, if "transposed onto the temporal order, this teaching of St. Paul is at the very heart of utopia [...] to strip off the old self means to rid man of those things that keep him from 'full being,' the things that make him selfish, apathetic, and unhappy."[28] This stripping, in turn, could be interpreted as an essentially anorexic gesture, since it implies an en-lightenment and a reduction. Whoever manages to succeed in this effort may achieve that a-topic dimension that is a sort of degree zero, from which it is possible to start over. In the utopian/dystopian context we are dealing with, therefore, any kind of metaphorical stripping (such as the poet's taking his life away) needs ultimately to be interpreted in the light of an allegorical critique of the superimposed, "tiresome vestments" of history and an implicit proclamation of the necessity of shedding them in order to change things, to achieve a new, creative status. In short, at the conclusion of the play, what the victorious self-effacing gesture of the poet (together with the vampire's refusal to eat any longer) seems to suggest, is that an alternative sickness, this time coming out of an anorexic-utopian embryo, may eventually and

28 In: *Utopia: The Search for the Ideal Society in the Western World*, Edited by Roland Schaer, Gregory Claeys, and Lyman Tower Sargent (Oxford: Oxford University Press, 2000), 319.

paradoxically be able (even though in an undetermined future) to affect the otherwise thoroughly dyspepsic-dystopian scenario of *Le Roi Bombance.*

It is perhaps an unavoidable consequence of the wealth of the texts I am dealing with that I began this section by emphasizing the dystopian elements in *Lettres de Malaisie* in order to try to demonstrate its potential similarity to the counter-utopian situation in *Le Roi Bombance*, and I am now concluding it by suggesting instead that an embryonic utopia may, after all, be (also) present in Marinetti's play. It seems equally evident, however, that the relationship between these two works may emerge stronger rather than weaker from the kind of reading I propose. Krishan Kumar may well summarize the core of such a relationship when he remarks that, at the end of the nineteenth and at the beginning of the twentieth "The line between utopia and anti-utopia is blurred; the possibility seems to exist for utopia to degenerate into anti-utopia and, correspondingly, for anti-utopia to modulate into utopia" (in Schaer, 262). It seems to me that Marinetti's and Adam's texts walk together along such a line. If Adam's *Lettres de Malaisie* may be defined as an apparent utopia that hides a dystopia, *Le Roi Bombance*, in this perspective, could be comparatively and reversibly defined as an evident dystopia that hides the potentiality of utopia.

That the kind of happiness that the *Lettres de Malaisie* describes is one achieved only by a comparable extinction (or stripping) of individual identity may thus be another sign of the invisible thread that links these two authors and their respective texts. It is, in fact, only through the disappearance of one's distinctive self and its subsequent absorption into the collective, social universe that the utopian *bonheur* seems to be achieved in Adam's *Malaisie*. It is only when such a reduced status is achieved that one can euphorically feel the "immense courant de la Gnose" and the "jous-sance intellectuelle" that derives from the "harmonie des forces" (183–4) (Minerva 1995, 186).

Together with this similarity, however, one needs to note a crucial difference from the situation in *Le Roi Bombance*. For Adam, a "dissolution of the self" may positively achieve a *bonheur*, but one partial and incomplete as it ultimately characterizes an ataraxic societal model constituted by homogeneous, bored, and indifferent human beings: "numéros sans caractère, sans passion [...] De purs esprits" (176), or, to put it in another way, not exactly a utopia (Minerva 1995, 87). Even when plenty of food is enjoyed – to conclude by re-introducing the gastronomical question – one in which "Celui qui a faim entre dans un réfectoire et mange à son appétit, boit à sa soif, autant qu'il le veut, quarante fois par jour s'il lui plaît." (47–8), and one in which "nos estomacs deviennent les plus choyés du monde" (133), happiness and freedom are problematic. Adam, of course, never suggests (or allows us to infer) that a "positively disturbing" anorexic behavior may improve things in this area,[29] even though he clearly implies that the complete absence of conflicts (or "malaises") is equally unsatisfactory. It is also more and more evident, in addition, that whenever Marinetti alludes to utopia he is already – as he, in general, shall always be – thinking of an aesthetic one, while Adam's vision and correspondent solution seems to still refer to a more concrete socio-political sphere. As a consequence, if the self-effacing "aesthetic-spiritual" option may be (paradoxically) satisfactory for the former, ending up a "pur esprit" cannot be ultimately enough for the latter.

Though in different measures, notions such as "victory", "hope" and "dream" are not alien to these texts, it remains true that universal happiness does not belong here. "Le Bonheur est ailleurs!" exclaims Sainte Pourriture in her final intervention in *Le Roi Bombance* while, in *Lettres de Malaisie*, the diplomat repeatedly wonders what happened to "la libertè" and "la personnalité" in the Malayan colony. Where to find such values and how to look for them are questions that haunted, in different ways, both Marinetti and his *maître* throughout their artistic careers. The former needed to create an avant-garde movement in his search for an answer; the latter kept producing novels and essays in which the problem of the origin and destiny of man was a constant thematic motif.

A renewed attention to the relevant alimentary dimension in *Lettres de Malaisie* and *Le Roi Bombance*, and the discussion of the dystopian/

29 Unless one counts the final infatuation (*amour*) between the "ami espagnol" and his guide, Pythie, and the subsequent change that may come out of the cancellation of each one into the other: "Comme tu m'as changée, toi, toi! […] Toi que me fais l'ennemie de mes espoirs, de mes croyances, de tout ce qui constituait mon être […] Tu es là; je n'existe plus qu'en toi … Oh, tes lèvres et la force de tes yeux!" (211).

utopian mode which informs both texts suggest that reading one against the other enriches their respective interpretations. The establishment of an intertextual relationship indirectly reaffirms Marinetti's early connections with *symbolisme* and a related strain of French socialist-anarchic thought. Second, the individuation of Adam's genealogical presence in the play may conclusively add a couple of creative definitions (perhaps "*comédystopie*" or "*tragédutopie*") to its classic ones of *tragédie satirique* or *hilarotragédie*. And, finally, this link with the general discourse about utopia at the *fin de siècle* may eventually allow me, in a future projection,[30] to add one more explanation to the otherwise apparently absurd and contradictory faith in progress, change, the power of machines, and the future that Marinetti would proclaim only a few years after the apparent pessimistic and static outlook of this early work. Well-known justifications such as the normal presence of contradictions in the personality of the founder of futurism, or the fact that "il desiderio distruttivo, nihilistico del Futurismo trova le sue radici in questa [i.e., the one in *Le Roi Bombance*] visione totalmente negativa della storia" (Eruli 1992, 159), in fact, could be perhaps joined by one more consideration. It is precisely from the recuperation and evolution of the individuated embryonic utopian vision in *Le Roi Bombance*, from the transformation of a primitive, messy, organic, body-centered, static dystopia which negates progress into a hygienic, modern, mechanistic utopia which attempts to free (or "relieve") the future from the past, that the ideal of futurism would soon come to light. Appropriately, that shall be a (utopian) time when the body has managed to efface its organicity, has escaped its dysfunctional, dyspepsic *malaises*.

By following the same controversial and blurred line which Marinetti's and Adam's texts have so far individuated, Sainte Pourriture, the "force absolue et unique qui demeure toujours identique à elle-même!" whose "triple force" consists in "Création! Destruction! ... et Régénération!" (260), shall thus – metaphorically speaking – be able to leave the scene to

30 And here Tamburri's notion of "retrolector" comes in handy. See Anthony Tamburri, *Semiotics of Rereading* (Madison, Teaneck, Farleigh-Dickinson University Press, 2003).

the other "absolute force," arguably her "updated sister" and much more frightening complemental ghost: the similarly equipped "Guerra, sola igiene del mondo!" which became one of the defining, dystopian events of the decade following the composition of *Le Roi Bombance*, and at the same time, among the crucial markers of the futurist utopia.

The following section of this chapter will thus continue to reflect on the characteristics and genealogies of this utopia as it manifests itself in the quintessential futurist alimentary text: Marinetti's and Fillìa's *The Futurist Cookbook*.

1.2 Past Ingredients for Future Foods: Traces of Fourier's "Gastrosophie" in *The Futurist Cookbook*

> This futurist cuisine of ours [...] wants to finally create a harmony between men's palate and their life of today and tomorrow.
>
> — F.T. MARINETTI AND FILLÌA, *The Futurist Cookbook*

> Gastronomy [...] becomes in Harmony a science of high social politics [...] in other words, Gastrosophie, high gastronomic wisdom, a profound and sublime theory of social equilibrium.
>
> — CHARLES FOURIER, *Théorie de l'unité universelle*

Futurism and Utopia

While this study may be the first to examine F.T. Marinetti's connection with Paul Adam's socialist utopian work, several scholars have already addressed the presence of utopian trends within futurism and, more particularly, Marinetti's familiarity with the precursors of European social-anarchic-libertarian thought (Stirner, Sorel, Saint-Simon, Proudhon). Only very few however, have expanded the subject beyond the discussion of the canonical writings in which futurism's utopian vision emerges more clearly (i.e., *War, the World's Only Hygiene*, 1915; *Futurist Democracy*, 1919; *Beyond Communism*, 1920). The objective of this section is to undertake such an expansion in order to suggest an interpretation of Marinetti and Fillìa's *The Futurist Cookbook* (1932),[31] not only as a work where the borders of art and life constantly merge, but as an essentially utopian text. In particular, this expansion may reveal interesting connections with the original "gastrosophic" theory on the role of meals and gastronomy in society, developed by Charles Fourier (1772–1837), one of the founding figures of the French utopian tradition at the cusp of the eighteenth and nineteenth centuries, not to mention his being a forefather of anarchism and a "patriarch of socialism."[32]

My goal is a comparative close reading of the selected texts that investigates two promising yet underdeveloped intuitions of Claudia Salaris in her study *Cibo futurista* on the futurist culinary tradition.[33] The first

31 F.T. Marinetti and Fillìa, *La cucina futurista* (1932) (Milano: Longanesi, 1986). Let me immediately clarify that, although Fillìa (Luigi Colombo, 1904–1936), the book's co-author, is himself a figure worthy of note, here I intend to focus only on Marinetti. Most of the sections I shall consider ("Against pastasciutta," "Invitation to Chemistry") display Marinetti's signature at their end and propose the overall ideology that Marinetti had already divulgated in several popular publications of the period. Fillìa's contribution to the book, probably also because he was a painter more than a writer, is mostly traceable in its conclusive, more visual sections, with which I am not specifically concerned here ("Determining Futurist Meals" and "Futurist Formulary for Restaurants and *Quisibeve*"). All the English translations provided are mine. See also *The Futurist Cookbook*, translation by Suzanne Brill (London: Trefoil Publications, Ltd. 1989).

32 Mirella Larizza, *Fourier*, ed. M. Ceretta, introd. M. Moneti Codignola (Firenze: Olschki, 2002), 21.

33 Claudia Salaris, *Cibo futurista: Dalla cucina nell'arte all'arte in cucina* (Roma: Stampa Alternativa, 2000), 8. Salaris had already gone over the history of the relationship of futurism and food. She observes that the manifesto "La cuisine futuriste," signed by the French cook Jules Maincave, but already bearing the evident influence of Marinetti, was first published in French in the journal *Fantasio* (1913). Following the Futurist gala dinners and events in the 1920s and a renewed interest in culinary matters in Italy in the 1930s, Marinetti and Fillìa subsequently re-presented the text to an Italian audience in 1930, via the gastronomical magazine *La Cucina Italiana*.

intuition is that "a certain influence of Fourier's utopia may be found also in Marinetti's theories," and the second, that both Fourier and Marinetti had a similar understanding of the social function of food and of the need to increase the amount of time dedicated to pleasurable activities, including cooking.

Both the significance and the critical benefits deriving from a deeper observation of Marinetti's alleged connection to Fourier – even if mostly limited to the gastronomical sphere – should not be difficult to define. These benefits may extend beyond the reassertion of the centrality of food aesthetics in the modernist avant-garde or that culinary issues are an overlooked part of utopian thinking. This analysis may point out another instance of futurism's controversial relationship to tradition, as exemplified through this additional re-appropriation, re-adaptation and, to quote Cinzia Sartini Blum, "recycling of the past." (Sartini Blum, 137). On the other hand, Marinetti's link to Fourier and a renewed attention to his early exposure to a French social-utopian environment may allow to reconsider the valid, if somewhat rigid, distinction in the standard critical perception of his work, between the early pessimistic, nihilistic, pre-futurist phase and the later, post-World War I futurist elaboration of a more coherent vision of an ideal society.[34]

The book itself, as a heterogeneous ensemble of documents (manifestos, journal articles about the current "food campaign," descriptions of past culinary events, fictional narratives and extravagant food recipes), was originally published in 1932 by the Milanese publisher Sonzogno.

34 Günter Berghaus notes that during his Parisian stay, Marinetti regularly visited the Abbaye de Créteil, a utopian community near Paris founded by Georges Duhamen and Charles Vildrac. Günter Berghaus, *Futurism and Politics: Between Anarchist Rebellion and Fascist Reaction, 1909–1944* (Providence, RI: Berghahn Books, 1996), 36–7. Moreover, even earlier, in the years 1888–93, while a young student at the Jesuit college Saint Francois Xavier in Alexandria in Egypt, Marinetti notoriously followed a "a solid French scholastic training," one which may well have included a thinker of Fourier's caliber. See F.T. Marinetti, *Teoria e invenzione futurista* [*TIF*], Luciano De Maria, ed (Milano: Mondadori, 1990), ciii. Interestingly, in this same period, he writes at least one essay on Henri de Saint-Simon (1760–1825), the French philosopher and social innovator, whose name is frequently associated with Fourier's

Matei Calinescu's comments about the lack of utopias and the omnipresent destructive nihilism that characterizes modernism and the cultural avant-gardes are well known.[35] However, a substantial utopian component, although not always immediately visible and veering more towards the individual-artistic and the literary than to the social-political side, has always been an integral part of Marinetti's thought, even though his pre-futurist works may generally indicate the contrary. Not surprisingly, Luciano De Maria wrote about the "afflato prometeico" and the "anelito di rinnovamento totale" that characterized Marinetti's ideology from the beginning of his literary career.[36]

As previously noted, the *fin de siècle* and the first decades of the twentieth century were a period when the line between utopia and anti-utopia was blurred, when utopia could degenerate into anti-utopia and vice versa. In this light, the end-product of Marinetti and Fillìa's *The Futurist Cookbook*, with its "clean," stylized, designer-made recipes aimed at improving the "sanity, agility, freshness of Italian intellectualism" (56) could be considered as the "logical" development of the culinary dystopia displayed in the nightmarish, "hard to digest" and dyspeptic scenario of *Le Roi Bombance* in which, as we have seen, utopian discourses are systematically reduced to alimentary and digestive matters in a fatal, downward spiral, apparently to be remedied by poetry and art alone. These two otherwise very diverse and temporally distant texts by Marinetti could then be viewed as part of a developing cultural program and ideology in which notions of social happiness and harmony, politics, aesthetics, human health, and progress keep recurring and are held together by an alimentary discourse infused either by a dystopian or utopian rhetoric (or both). Let me emphasize that only four years separate the apparently nihilistic and anti-utopian *Le Roi Bombance* from the founding 1909 manifesto that expressed "the utopian

---

under the common label of reformer and socialist utopian. See Francois Dagonet, *Trois philosophies revisiteés: Saint Simon, Proudhon, Fourier* (Hildesheim: G. Olms, 1997).

35 Matei Calinescu, *Five Faces of Modernity: Modernism, Avant-garde, Decadence, Kitsch, Postmodernism* (Durham, NC: Duke University Press, 1987).

36 In Marinetti, *TIF*, xxxviii, xlii.

belief in a new art of the future that [...] would no longer serve as an escape from life but directly engage and transform it."[37] The "'anarchia dei perfezionamenti [...] in cui si condensa l'essenza più pura e preziosa dell'utopia sociale futurista" (*TIF*, xliv) derives from an embryonic utopian germ that had always been present in Marinetti's thought.

Thus, with characteristic opportunism and time-awareness, Marinetti may have eventually modulated the pessimistic, body-centered, progress-negating, counter-utopia represented in his 1905 drama into a modern, mechanized, technologically obsessed, and nature-taming "gastro-utopia," which, interestingly, through a "digestive" gesture, had as one of its main targets the elimination of the past. This new era, as *The Futurist Cookbook* itself suggests, would be "spiritual" and "harmonic," a time when the body would shed its natural, dysfunctional, dyspeptic *malaises* and morph into a lean, agile, efficient machine, ideally able to function altogether without food (or love).

The Futurist Cookbook needs also to be contextualized within a cultural and political project aimed at creating masses of healthier and stronger super-citizens, as demanded by fascist Italy, and that continues futurism's iconoclastic rejection of the past and tradition. Its proposal to abolish the heavy *pastasciutta*, for example, was certainly motivated less by alimentary issues than by an attempt to dismiss the whole passéist Italian culinary establishment. Besides its anti-traditionalism and politically motivated, autarchic scope,[38] such a refusal is probably the most famous and ostentatious point of this utopian, "enlightening" project:[39]

37 Anne Bowler, "Italian Futurism and Fascism," *Theory and Society* 20.6 (December 1991), 785.

38 At the same time, Carol Helstoski observes that "*The Futurist Cookbook* appears to have little to do with fascist policies, which aimed to conserve available food supplies." Carol Helstosky, "Recipe for the Nation: Reading Italian History through *La scienza in cucina* and *La cucina futurista*," *Food & Foodways* 11 (2003), 129. For a discussion of Marinetti's rejection of pasta, see also Michel Delville, "Contro la pastasciutta: F.T. Marinetti's Futurist Lunch," *Interval(le)s* 1.2 (2007), 14–24.

39 It should also be mentioned that the allusion to the lightness and the spiritual and physical flexibility that Italians must possess anticipates the nutritional components of "naturism." This was another utopian strain within Futurism, interconnected with

> The futurist culinary revolution [...] proposes [...] to radically modify the alimentation of our race, strengthening it, making it dynamic and spiritualizing it with very new foods. [...] This futurist cuisine of ours, regulated like the engine of a hydroplane for high speeds [...] intends to finally create a harmony between men's palate and their life of today and tomorrow [...]. We believe it is necessary, first of all: a) to abolish pastasciutta, the absurd Italian gastronomic religion (28).

If, as Michel Onfray observes, social revolution for Marx was the way to liberate man, for Marinetti liberation and the promise of an improved future begins with an alimentary revolution: a modified, re-organized cuisine becomes a way to revolutionize society and reality as a whole and to create a new, better life.[40]

Fourier's Utopian Vision

Having hinted at the complex relationship between futurism and utopianism and suggested *The Futurist Cookbook*'s own utopian dimension, with its aspiration to be a "cook-book for the kitchens of the future," I now turn to consider Fourier's utopian vision and the significance of the alimentary discourse in his writings.

Fourier's culinary-philosophical concept of *gastrosophie* is just a fragment of a much larger system of thought aimed at the radical redefinition of reality through a revolutionary project of comprehensive social reform and liberation from the contradictions of a rising capitalism. His main objective was to achieve a state of universal harmony (*Harmonie*) in a fairer, alternative, ideal society resolutely set against the proto-capitalist, bourgeois society of post-revolutionary (and *post-philosophes*) France at

---

fascist ideals of physical culture and autarchy. Its manifesto by Marinetti and Ginna ("Il naturismo futurista," 1934) proposed, among other things, an alimentation that "makes the spirit nimble" (Salaris, *Cibo futurista*, 77).

40 Michel Onfray, *Le ventre des philosophes: Critique de la raison diététique* (Paris: Bernard Grasset, 1989), 154.

the beginning of the nineteenth century.[41] His thoughts about gastronomy and *gourmandise* are not gathered into a single work but need to be individuated and extracted out of a corpus of several volumes (from the more than twelve that constitute his *oeuvre*) published between 1808 and 1849.[42]

Fourier intends his theory to be a powerful vehicle for regeneration by assigning – even to culinary matters – society-changing values similar to those usually assigned to the economic, moral, social, or political spheres. He seeks to transform civilization's fundamental problems (starting with socio-economic inequality, freedom from oppressive "unattractive" labor, and class differences), with a condition of universal happiness. Fourier's scholars commonly assign crucial roles to his revaluation and liberation of the human passions and desires, which he understood as a necessary step towards the elimination of false and repressive moral values; his critique of commerce, modern industry and technology, which, by introducing machines, have dehumanized labor and increased unemployment; his cyclical, cosmic understanding of human history, oscillating between progress and regression; his concept of a non-subjugating, pacific and pleasurable relationship between man and nature; and, last but not least, his cult of artistic beauty and the importance given to play and aesthetic activities.

As this necessarily brief summary of Fourier's multifaceted theory shows, the differences in context, scope, and underlying ideology between his politically and economically motivated effort to imagine an alternative society and Marinetti's mostly a-political utopian vision, aiming to substitute class struggle with a mystic struggle that disregards economic issues and

41 Such a society was supposed to be organized in groups of people ("Phalanxes") hosted in a complex of buildings ("Phalansteries"). I cannot help noticing that in *Futurist Democracy* Marinetti writes about instituting an "Eccitatorio" made up of a group of young members that should replace the "decrepita e ingombrante instiuzione del senato" (*TIF*, 416).

42 In particular, Charles Fourier, *Théorie des Quatre Mouvements* (*TQM*) (Paris: J. Jacques Pauvert, Éditeur, 1967); *Le Nouveau Monde amoureux* (*NMA*) (Paris: Slatkine, 1979); *Théorie de l'Unité Universelle* (*TUU*) IV (Paris: Éditions Anthropos, 1966); *Le Nouveau Monde Industriel et Sociétaire* (*NMIS*) VI (Paris: Éditions Anthropos, 1966).

to privilege the role of art in solving social problems, emerge quite clearly. If we then add to these Fourier's truly avant-garde (and *avant-le-temps*) condemnation of women's subordination,[43] his dismissive attitude towards technology and machines, or his belief in the function of "peaceful armies" and in the false myth of progress, it is evident that the cultural and ideological distance between the French philosopher and many of Marinetti's and futurism's most notorious positions could not be wider. However, one should not categorically exclude the possibility that Marinetti may have come across and appropriated or adapted scattered parts of Fourier's theories to his own purposes, or that these two visionaries may have shared substantially similar attitudes towards reality and the ways of modifying it.

It is also likely that any comparison between the Fourierist and futurist utopias may be more profitable if it emphasizes the aesthetic over the strictly political aspect. In this sense, Italo Calvino's 1971 introduction to a selection of the French philosopher's *oeuvre* is particularly encouraging for my purpose, since he writes that Fourier "establishes a relationship of affinity with the work of artists and poets, with their manipulation of the linguistic and mythical matter and their hope of managing to 'change life' through it." Calvino then warns us that we should distinguish between, on the one hand, a "reading that makes an immediate, political use of Fourier's text" and, on the other, "an attitude that appreciates its outdatedness and impracticability, [that sees it] as a challenge and perennial provocation, as a stimulus to go beyond the '[things]-in-existence,' to measure oneself with alternative worlds, in a dimension that remains literary and cannot

43 Fourier's position on questions of gender and sexuality was incredibly open-minded and advanced for his (but also our) times. He was not only "an early advocate of women's liberation" (James W. Brown, "Alimentary Discourse in Nineteenth-Century Social Theory: Pierre Leroux, Etienne Cabet and Charles Fourier," *Dalhousie French Studies* 11 [1986], 90), but he believed, for example, that in Harmony, every man and woman had the right to have a satisfying minimum of sexual pleasure, and he promoted free love, polygamy and "the universal gratification of sexual desire, even when desire did not have 'normal' heterosexual congress as its object." J. Beecher and R. Bienvenu, *The Utopian Vision of Charles Fourier* (Boston, MA: Beacon Press, 1971), 60.

become a program of political action."[44] It is, therefore, with Calvino's critical attitude in mind that I approach Fourier's and Marinetti's respective alimentary discourses.

Marinetti and Fourier: Differences and Convergences

An initial, bridging step is to observe that both figures defy easy pigeonholing. Much like Marinetti's own theorizing, Fourier's is not without ambivalences and, as such, has elicited and benefited from contrasting critical interpretations.[45] For example, in "Paris, Capital of the Nineteenth Century" ("Fourier, or the Arcades"), when Walter Benjamin remarks that "the secret cue for the Fourierist utopia is the advent of machines," or that "in Fourier, technology appears as the spark that ignites the powder of nature," he also indicates that the technological revolution taking place during Fourier's lifetime left quite significant traces in him as well.[46] By taking Benjamin's insights literally, Mirella Larizza goes even further in challenging Fourier's supposed hostility to progress and technology when she asserts that he "absolutely does not turn his shoulders to technology and human conquests but, rather, he thinks of the relationship between man and nature in terms of subjugation of the latter, so that human needs may be satisfied."[47]

Similarly, Carmelo Romeo identifies even more specific ambivalences when he observes first that, while Fourier criticized science and industry,

44 Charles Fourier, *Teoria dei quattro movimenti, Il nuovo mondo amoroso e altri scritti sul lavoro, l'educazione, l'architettura nella società di Armonia*, Scelta e introduzione di I. Calvino (Torino: Einaudi, 1971), xxix, xix (my translation).

45 De Maria first noted that "l'esame del futurismo è l'esame delle sue contraddizioni" (Marinetti, *TIF*, xlvii) and, more recently, Sartini Blum added that "futurism combines a penchant for outrageous utopian visions with a persistent interest in the practical, efficient aspects of modern life" (Sartini Blum, 18).

46 Walter Benjamin, *The Arcades Project*, ed. Roy Tiedemann, trans. Howard Eiland and Kevin McLaughlin (Cambridge, MA: The Belknap Press, 1999), 16–17.

47 Larizza, *Fourier*, xxx.

at the same time he "reserves for the most deserving members of the Phalanstery, that is the industrialists and the technicians, the prerogative to occupy a spot in the Aeropagus [a supreme council which supervised the organization of work]," and, second, that since he does not actually choose between "the plan of slow evolution (in order to get out of social chaos) and the one based on force ... his alleged pacifism needs to be reconsidered."[48]

Not surprisingly, similar ambivalent interpretations may be made in relation to Fourier's and Marinetti's writings about food. The former's "gastrosophic" theory, for example, has been defined as a true "socialist meal theor[y],"[49] since it was aimed at improving the eating habits and feeding conditions of all men, regardless of social class. Marinetti's own position, on the other hand, has not only been said to serve the alimentary policy of the fascist regime but to coincide with an elitist and aristocratic culinary aesthetic,[50] which, again, would appear to be in direct contrast to the social-oriented concerns of his illustrious predecessor.

However, when the figure of the *gastrosophe* is considered, one realizes that this "he" or "she" is a special and unique person who, only after rising above the masses and completing decades of training in the culinary art, may preside at the top as leader of Fourier's gastronomical project.[51] Such a figure evidently privileges individualism over collectivism, difference over uniformity. In other words, the *gastrosophe* could be rightfully considered as belonging to a limited, and therefore quite elitist, aristocracy of food experts in society.

Conversely, as Onfray provocatively observes, Marinetti's theoretical stance in *The Futurist cookbook* is in some ways closer to Hegel than to Nietzsche. With this, he means that Marinetti, by operating a "critique of the individual faculty to judge in order to promote the principle of a social judgment, for the general interest," is more interested in the ultimate

48 Carmelo Romeo, *Introduzione a Charles Fourier* (Messina: Samperi, 1983), 78.

49 Brown, "Alimentary Discourse in Nineteenth-Century Social Theory." 74.

50 Onfray, *Le ventre*, 162.

51 According to Fourier, one can become a *gastrosophe* only after reaching the age of eighty years.

good of the many than merely that of the individual.[52] In other words, the aristocratization of the crowd, the transformation of the mass into an elite group, or in Marinetti's words, a "proletariato dei geniali" (*TIF*, 405), possibly willing to adopt and enjoy the new diet, is also included in the main objectives of the alimentary revolution proposed by the founder of futurism.

A text like "Il proletariato dei geniali" (*TIF*, 405) reminds us that Marinetti's writings often oscillate between expressions of elitism and celebrations of popular, revolutionary crowds. Other statements such as "Abbasso l'eguaglianza!" "Scatenate dovunque e esasperate l'originalità individuale," or "aumentate le ineguaglianze umane" ("Inegualismo e Artecrazia" 1922; *TIF*, 551, 553) coexist with more democratic and socially engaged pronunciations such as "Tutti i cittadini avranno il diritto di esporre [art works] gratuitamente," or "bisogna imporre al più presto l'azionariato sociale cioè: la partecipazione degli operai alle imprese" ("Democrazia futurista," 1919; *TIF*, 409, 421).

Obviously, these remarks do not suddenly transform Marinetti into a socialist (or Fourier into an elitist), nor can they eliminate the perception of the provocative playfulness of the former's cultural operation, his familiar mixture of the political and the aesthetic *vis-à-vis* the apparently more serious social concerns of the French philosopher. At the same time, however, they may question the absoluteness of Marinetti's utopian elitism in contrast to Fourier's populism. In short, when Marinetti hopes that such a "proletariat of men of genius" may reach the stage of minimum manual work together with the maximum salary, which, "senza diminuire la produzione, potranno dare a tutte le intelligenze la libertà di pensare, di creare, di godere artisticamente" (*TIF*, 486), he does not seem to be far from Fourier's idea of "attractive work" that is justly rewarded, varied, playful, and productive at the same time.

It is, therefore, within this sporadically convergent intellectual and imaginative framework that Fourier and Marinetti approach the issues of gastronomy and culinary art and consider them as crucial elements of an

52 Onfray, *Le ventre*, 157–8.

ideal society. They are "the foundations of the building," to quote the former, able to affect and therefore revolutionize reality through a series of changes in the ways that, respectively, the French and the Italian behave in matters of nutrition. Both writers expand and underline the importance that food can have in society and the ability it can have to transform and improve an unsatisfactory socio-economic situation. Fourier's socio-gastronomical vision appears to be a significant theoretical precedent and a striking source of images for Marinetti (even when compared with Anthelme Brillat-Savarin's *Physiology of Taste* [1826]), not to be overlooked when dealing with *The Futurist Cookbook*.[53]

A Closer, Comparative Look at the Texts

> [Note] the fashion of that rancid glue called "vermicelli," that has become the general soup in Paris, because it makes the grocer rich and saves time for the cook.
>
> — FOURIER, *Le Nouveau Monde Industriel et Sociétaire*

This criticism of *vermicelli* could perhaps have legitimately been written during one of his numerous Parisian stays, by the same figure who virulently attacked *pastasciutta* for its undesired moral-physiological effects: "weariness," "pessimism," negative socio-political impact, "nostalgic inactivity and neutralism" (Marinetti, *Cookbook*, 29) and historic-economic inappropriateness. These words, however, are not Marinetti's but Fourier's, and in the rest of chapter twenty-six of his *Le Nouveau Monde Industriel et Sociétaire* (1845) he goes on to discuss the negative consequences of privileging a "mercantile gastronomy" in France (in Paris, in particular, as an exemplary part for the whole), and criticizes contemporary national

53 Salaris observes that Marinetti most likely consulted Anthelme Brillat-Savarin's *Physiology of Taste* (1826), which anticipates his interest in the integration, concatenation, and multiplication of the five senses, the introduction of chemistry into alimentary science, and the notion of the "edible landscape," one of the strongholds of futurist cuisine (Salaris, 9–21).

culinary practices and attitudes: "we observe that Paris, that is the center of fine arts, is also the center of bad taste in gastronomy. Parisians indifferently consume the good and the bad" (*NMIS*, 255). This quotation needs to be read as part of Fourier's criticism of commerce, which he saw as a "sterile, useless, parasitic function." While in Paris, he finds the abundance of fine arts does not correspond to an abundance of good food, and civil society as a whole is compared elsewhere to a "badly laid table."[54] An ideal of equilibrium and symmetry has been lost and needs to be re-established. In this sense, the opposition between "the center of fine arts" and "the center of bad taste in gastronomy" may suggest that for him the aesthetic and the gastronomic (not to mention the "economic") are, or should be, intimately linked. Even more significantly, the two should not be at the service of utilitarian and mercantile purposes. In an ideal society, artistic beauty and the cultivation of the fine arts should be inseparable from good nutritional habits and the cultivation of fine taste in food. One should not exist without the other and both, in turn, should enrich, in a metaphorical rather than strictly materialistic sense, all people rather than being instruments of financial gain for a single subject or professional category. Food and meals, together with their other, fundamentally socio-political roles, are thus elevated from their elementary, life-supporting functionality to a higher sphere in which they fulfill not only a socio-economic and alimentary need but provide aesthetic pleasure.

As Fourier suggests in the paragraphs on "La cuisine en harmonie" in *Le Nouveau monde amoreux*, the "wise" ability to prepare (and then discuss) a variety of excellent courses and present them in an aesthetically appropriate manner should be one of the main objectives of all citizens/cooks of *Harmonie*: "Each [person in Harmony] claims to excel in the preparation of some variety of food and contributes to the work in the kitchen on the day when such food becomes object of discussion and feast." When this point is reached, Fourier adds, "the cuisine in this new order is more or less the science of the whole world" (*NMA*, 131). Contemporary Parisians, however, because of their lack of gastronomic and, implicitly, aesthetic

54 Larizza, 13/17.

judgment, are said to be "vandals with regards to gastronomy." Their eating pasta (*vermicelli*) is one manifestation of their poor culinary education and dubious taste. Two more foolish, despicable deficiencies (or *gastro-âneries*, as he calls them) are pointed out in a digressive footnote. The first is that Parisian bakers and pastry makers have the self-serving, "monstrous" habit of half-baking their preparations so that they can be preserved longer and, in case of a sales-crisis, be re-sold the following day to the unknowing consumer. The second has to do with the Parisians' infatuation with *le genre anglais*, the widespread "anglomanias" that range from the fashion of eating rare meat with the help of almost impossible-to-hold, backward-bent forks to that of substituting the good, healthy foods of one's own native country with foreign, "base stuff' such as tea. Tea, he continues, is something that the British are "forced" to drink because they do not have their own good wine or good fruit, which they can only purchase at an enormous expense.

These descriptions and negative assessments of contemporary French food practices are, of course, in sharp contrast to what will happen once Fourier's ideal society and his harmonious system become functional. And that shall happen in the blissful future time and place where *vermicelli* is banished and disappears from menus, pastries are fully baked, and people eat the food produced and prepared in their own country rather than abroad. More importantly, this will be a time when "mercantile cunning" can no longer influence the public's taste and the gastronomic institution is responsible for the functioning of the economic system and for imaginatively regulating the two recurring cornerstones of Fourier's whole philosophical system – the passions and desires: "gastronomy [...] in Harmony will be imagined as the main source for balancing the passions" (*NMIS*, 258).

This glance at Fourier's alimentary discourse may already evoke a series of corresponding images in *The Futurist Cookbook*. Consider, for example, besides Marinetti's own rejection of *pastasciutta*, his proposal to ban the use of forks and knives, his nationalistic hatred of foreign foods and his general xenophobic attitude. Of course, Marinetti's pasta rejection has a different motivation; his banning of cutlery is explained by his search for a "tactile pre-labial pleasure" (*Cookbook*, 32) more than by ergonomic considerations, and his fascist-inspired xenophobia goes well beyond the alimentary sphere. However, despite these differences, the recurrence of this kind of imagery and argument seems to bode well for the existence of other, less superficial parallelisms.

In line with the parameters of most utopian texts, *The Futurist Cookbook* begins with a negative assessment of a past or contemporary situation. In this case, the starting point is the state of culinary art in Italy: "Men, until now, have fed like ants, mice, cats and oxen." The eminently artistic and aesthetic quality that the new *cucina* must have is then immediately pointed out: "With us futurists the art of eating is born. As with all the arts, it excludes plagiarism and demands creative originality" (6) and, together with it, the objective of replacing courses that are expensive for the consumer, trivial and repetitive, with dishes which are cost-effective and privilege creativity and imagination: "very new foods in which experience, intelligence and fantasy economically substitute quantity, triviality, repetition and cost."

Fourier's search for combinations of foods which appeal to the palate and to the eye, as well as his original combination of the aesthetic with the economic, come easily to mind when one is faced with the new and imaginative foods of futurism, made more to be seen and discussed than eaten, and with Marinetti's own interpretation of the role of culinary art in the economy of the Italian nation. Additionally, the fact that the futurist aesthetization of the culinary experience is a necessary step to making this experience potentially accessible and useful to everyone may coherently recall Fourier's own version of "romantic socialism," which proposed the positive effects of the synergy between the flavorful and the beautiful for the cause of the social good. Marinetti's alimentary revolution, after all, must be made of ingredients that are ultimately useful to everyone: "intelligence" "fantasy" and "creative originality." For Fourier, instead, one should "free desires, let the instincts range free, authorize the imaginary to rule over reality, in a word, take one's desires as reality."[55] Besides the shared emphasis on creativity and imagination, it is precisely this question of desire – of being able to free, fuel, and satisfy one's desires – that seems to represent the most significant analogy and common concern for both authors.

55 Onfray, 107.

"How can one manage to have an appetite that perennially renews itself?" Fourier rhetorically asks in the chapter "Of Politics Applied to Appetite" (*NMA*, 133). His answer is multifaceted and elaborated throughout his work. Good food, variety, right timing between meals, good company and attention to quantity are, in short, the qualities he deems necessary to keep the appetite or, alternatively, the "alimentary desire" alive. The secret of harmonious politics, the way to achieve *le bonheur*, as he puts it, is first to be sought in the quality and variety of foods on one's table: "As far as the variety of dishes at their table, there will be at least thirty or forty of them, a third of which will be changed each day, with a dozen different kinds of drink at every meal" (*TQM*, 165). The "regime of variations" he proposes has, in turn, the extra benefit of speeding up the production of desire (so that one may look forward to the next meal in the day), accelerating digestion and, significantly, of not making one's body and spirit heavy: "variations in the alimentary system achieve[d] the objective of accelerating digestion, sustaining the appetite, repeating it ceaselessly" (*NMA*, 134). The daily boredom and tediousness that usually derive from traditional preparations and from not knowing how to arrange and organize a meal (in terms of the menu and the guests) will be avoided in Harmony. In this utopian context, the pleasures of the table will depend on the simultaneous presence of the right food and the right company: "the judicious mix of table companions, the art of varying and matching people, of making each day more interesting by unexpected and delicious encounters" (*TQM*, 168).

Fourier's gastronomic/existential question regarding the nature and the never-ending preservation of one's desire also haunts Marinetti, although he never posed it in such an explicit and direct fashion. His statement in *The Futurist Cookbook* about the introduction of a musical interval in-between courses so that it "functions to wipe out [an earlier] enjoyed taste [in order to] re-establish a tasting virginity" (33), suggests that Marinetti, like Fourier before him, is concerned with the possibility of perpetually hitting a sort of reload button, to renew and refresh forever one's desire and subsequent pleasure, in a sort of culinary eternal return.[56]

Being able to eternally "renew the appetite" or to re-establish an indefinitely renewable "tasting virginity" became, then, part of a similar rhetoric that betrays both its utopian trait and a dynamic and transformative understanding of the culinary microcosm and the world. In both cases, the systematic will to keep everyone's desire and pleasure alive becomes crucial to the transformation and improvement of society as a whole and, as such, represents another comparable point between the two alimentary discourses under examination. If it is true that for Fourier "the most basic pleasure is gastronomical fulfillment" and "man will eventually evolve to a state of Harmony by following his passions,"[57] Marinetti's project to substitute "crisis" with "harmony," "panic" with "optimism," and, as we shall see, even "death" with "life," with the help of food, originates in a "passionate" reinterpretation of traditional nutritional habits, with the aim of attaining the pleasurable and, if possible, simultaneous fulfillment of man's five senses.[58] Passion, imagination, beauty, creativity, desire, and pleasure thus become recurrent keywords in both contexts and the means of moving towards a universal harmony, be it in nineteenth-century France or twentieth-century Italy.

In the end, Marinetti, who is equally concerned with the renewal of pleasure and desire, answers Fourier's question in quite compatible tones. This said, an opposition unavoidably remains between the latter's understanding of (gastronomical) harmony as a socially-aware condition linked to notions of abundance and widespread access to food, and the former's emphasis on harmony as an essentially aesthetic construction, a status

56 Onfray significantly mentions the concept of "eternal return" when he writes on both Fourier and Marinetti: "La volonté fourieriste est la maintenance du désir dans son eternal retour," and then with regards to *La cucina futurista*: "La diététique comme révélateur de l'éternel retour" (Onfray, 123/177).

57 Brown, "Alimentary Discourse in Nineteenth-Century Social Theory," 82–3.

58 Fourier too deals with the importance of satisfying all the senses in Harmony (and not only taste).

paradoxically achieved by privileging the out-of-the-ordinary, dissonance, and surprise, relatively unconcerned with edibility or taste.

Desire and/or appetite must nonetheless be kept alive by emphasizing variety, by avoiding boring companions at the table ("abolition of eloquence and politics at the table"), and by looking for unique combinations of unusual ingredients ("abolition of traditional mixtures in order to experience all the new and apparently absurd mixtures") (*Cookbook*, 33/29). His "perfect meal" not only attempts to create "an original harmony of the table (glassware, china, decoration) with the flavors and colors of foods" but also by the timely introduction of the food (by alternating it with foods that will not be eaten), always in order to "favor curiosity, surprise and fantasy" (30–3). Even more important is the fact that such a perfect meal should be easy to digest, so that it may prevent Italians from becoming "cubic, bulky, leaded by an opaque and blind compactness" (27), and promote the lightness, "agility" and "dash" without which any attempt to spiritualize and "dynamize" the "race" would be doomed.

The Futurist Cookbook and Hygienism

If digestible foods and specific nutritional behaviors represent an additional path towards optimism, health, and salvation, the consequences of eating the wrong food, in contrast, can be quite dangerous. The events narrated in "A Meal that Prevented a Suicide," the short piece that opens *The Futurist Cookbook*, for example, repeat the characteristic avant-garde trope of cannibalism in order to explicitly allude to the life-saving and therapeutic effects of the right nutrition.[59] In short, Giulio Onesti's suicide is prevented thanks to Marinetti and his friends arriving at his Umbrian villa on Lake Trasimeno and cooking him a very special dish: an edible sculpture in the shape of a woman, prepared with a variety of "indispensable ingredients" (11). After this erotic and life-giving "edible sculpture" has been consumed

59 Marinetti knew both Francis Picabia's "Le manifeste cannibale" (1920) and Oswald de Andrade's "O manifesto antropofago" (1928).

by Giulio in an anthropophagic gesture ("he ate a sweet foot [...] he bit the thick heart from the hearts of pleasure [...] he ate the mammillary spheres (sources of) every maternal milk"), he feels completely regenerated, as if he were in a state of ecstasy: "he was at once empty, freed, void and full. Enjoying and enjoyed. Possessor and possessed. Unique and complete" (20). The mystical-erotic tones pervading this episode, the reference to the Christian Communion, in which, instead, the symbolic body of a male savior (the Host) is eaten and, finally, Marinetti's ambiguous attitude to the female that has to be objectified and eaten so as to cease being a threat to the male protagonist and his friends, are all obvious considerations here.[60]

For my purposes, however, I would especially like to emphasize the implicit curative qualities and prophylactic effects that this food possesses. This special meal, in other words, is what a person and, arguably, the utopian society Marinetti has in mind, needs in order to stay alive, maintain good health, and prevent negative ailments such as self-destructiveness, "depressing panic," and pessimism from spreading further. Different from the traditional "way of feeding that prepares to suicide" (10), the aesthetic-erotic nutrition and diet that Marinetti introduces in these pages and continues elaborating both in his descriptions of other "Decisive Futurist Meals" and his "formulary," simultaneously saves Giulio's life and contributes to his spiritual and physical health. That Marinetti prefers the denomination "formulas of futurist cuisine" to "recipes" seems to validate the hypothesis of considering them as unusual and idiosyncratic food combinations, on the one hand, but – following a tradition of linking gastronomy and medicine which originates in the Middle Ages and continues in the Renaissance – as remedial prescriptions for the body and the soul of men, on the other. The section entitled "Invitation to Chemistry," which argues in favor of "giv[ing] soon to the body the necessary calories through State nutritional equivalents, powder or pills, albuminoid compounds, synthetic fats and

60 I am grateful to Pierpaolo Antonello for drawing my attention to the fact that the reference to the breasts and maternal milk recuperates an image already present in the original "Futurist Manifesto," following the episode of the car accident: "Io gustai avidamente la tua melma fortificante, che mi ricordò la santa mammella nera della mia nutrice sudanese ..." (*TIF*, 9).

vitamins" (30), provides further corroboration of the association between the gastronomic and the medical spheres.[61]

Taking this argument one step further, *The Futurist Cookbook* can be read as a cross between a cookbook and a pharmacopoeia. A text, in other words, in which food is medicine and medicine is food and which, by dealing with human "hygiene" in one of its most common manifestations, inevitably contributes and belongs to the far-reaching debate on "hygienism" in Italy (and Europe) between the end of the nineteenth and the beginning of the twentieth century. As the multifaceted figure of Paolo Mantegazza (1831–1910), one of the main Italian protagonists in this debate, reminds us in his *Treatise on Hygiene* (1864), hygienism was "a social science which … draws light and strength from physiology, and extends a hand to political economy and law."[62] A "physiologist, anthropologist, ethnologist, pathologist […] hygienist […] and among the propagators in Italy of Darwinian evolutionism,"[63] Mantegazza also included nutrition in the myriad of his interests. He and his followers after World War I believed that certain changes in the alimentary habits of Italians could have a role in preventing disease, protecting public health, and improving social hygiene as whole.[64] But the implied notion that the social body could be made healthier signals also the continuity and the intersections of Mantegazza's and, by extension,

61 Re: "Invitation to Chemistry," Salaris chose to emphasize its affinities with Brillat-Savarin, as "another point of contact between *The Physiology of Taste* and *The Futurist Cookbook*" (Salaris, 10).

62 In Edwige Comoy Fusaro, "Réflexions sur le texte (pseudo-)littéraire chez Paolo Mantegazza: *Un giorno a Madera. Una pagina dell'igiene d'amore*," *Cahiers de narratologie* 14 (2008), http://revel.unice.fr/cnarra/document.html?id=543 [Last accessed October 2012].

63 Fusaro, "Réflexions sur le texte (pseudo-)littéraire chez Paolo Mantegazza," 3.

64 Claudio Pogliano notes that "Since 1931 Silvestro Baglioni – physiologist, eugenicist, sexologist – published a supplement to his journal *Fisiologia e Medicina* entitled 'Il problema alimentare. Chimica – Fisiologia – Patologia – Terapia.' All the Fascist 'battles' – for grain, milk, rice, grape – had an echo or were anticipated there, since it was certain that the science of alimentation was crucial for the relationship between the economy of the human body and that of the social one." Claudio Pogliano, "Scienza e stirpe: eugenica in Italia (1912–1939)," *Passato e presente* 5 (1984), 92–3.

Marinetti's hygienic utopia with the more extreme interpretations of the new science of eugenics (i.e., the issues of racial hygiene and the utopias of human breeding) which would later constitute such a crucial subtext of the Nazi-fascist ideology.

This said, if I have lingered on the episode of the failed suicide and the "salvific meal" narrated in *La cucina* and some of its potential ramifications, it is because it seems possible to detect similarities between it and Fourier's culinary philosophy. I am referring in particular to those passages dealing with *gastronomie hygiénique*, a notion that "joins culinary expertise with the medical and pharmaceutical sciences."[65] As a matter of fact, Fourier had already introduced the idea of food as *pharmakon* (in Greek, *both* "remedy" *and* "poison"),[66] and the theory about the beneficial digestive effects deriving from sexual intercourse. The notion of *médecine du goût*, for example, that he describes as "the line of antidotes (that are) pleasant to administer with each sickness" and as "a branch of the science named hygienic gastrosophy, a preservative and curative method at the same time" (*NMIS*, 260), brings together the medical and the alimentary spheres in a way that seems to generally anticipate the key creative ingredient in "A Meal that Prevented a Suicide."

Sexual-gastronomic remarks such as "a moderate coitus expands the soul and helps digestion" and "if he, before dinner, had possessed a woman of his taste, this exercise, taken with moderation, would have better prepared him for digestion" (*NMA*, 135), reduce women to the level of a *hors d'oeuvre* or an aperitif that has the practical effect of facilitating men's digestion, and thus suggest that Fourier too, at least in these circumstances, objectifies women, considers them as something to be consumed, and includes them among those foods that are good for men's health. Onfray observes that "the

65 Brown, "Alimentary", 89.

66 Brown also notes that "Fourier [...] allies gastronomy with pharmacy on the basis of certain transmutations which occur either during or [during] the mixing of chemical substances" (89). A similar alliance is certainly pursued in Marinetti's "Invitation to Chemistry."

essential criterion of the alimentary hygiene of Fourier is digestibility,"[67] and the same could well be applied to Marinetti's "alimentary hygiene."

After having mentioned the potentially comparable notion of "hygienic gastronomy" in these two writers, it is tempting to conclude my reflections with another, well-known instance in which the term "hygiene" recurs in Marinetti, though not in *The Futurist Cookbook*. I am obviously referring to his famous statement in the 1909 *Futurist Manifesto*: "Noi vogliamo glorificare la guerra – sola igiene del mondo" (*TIF*, 11). The apparently similar regenerative and hygienic power assigned by Marinetti to both food and war suggests that he saw a relationship between the two. A remark such as "Il cuore deve diventare in qualche modo, una specie di stomaco del cervello, che si empirà metodicamente perché lo spirito possa entrare in azione" (*TIF*, 300), which aptly appears in "L'uomo moltiplicato e il Regno della macchina," in *Guerra sola igiene del mondo* (1915), would seem to confirm this hypothesis.

Fourier, in turn, discusses at length the concept of a "gastronomic fight of numerous armies," which is necessary to "determine the perfection of inferior food" (*NMA*, 339). The selection of the best and most digestible foods in Harmony happens because of a (peaceful) alimentary war. That is, war in Fourier is one of the means that determines the excellence of the foods destined for the citizens of Harmony, of their level of alimentary hygiene.[68] A conveniently modified, entirely made-up statement such as "war is the only *alimentary* hygiene of the world" could perhaps convey the fascination that the relationship between hygiene and cuisine, on the one hand, and hygiene and war, on the other, had for Fourier and Marinetti. It is also necessary, however, to point out the difference between the actual degenerative, social dimension of the hygienic practices related to war, and the individualistic, playful, and peaceful activities of those related to Fourierist and futurist alimentary discourses. Rather than being a purifying bloodbath hygienically and eugenically producing a stronger, healthier race, war, in fact, was also increasingly perceived as "dysgenic," a sort of

67 Onfray, 122.
68 Onfray, 120.

"reverse selection" that, by causing the deaths of the younger and stronger and leaving the survivors traumatized, hungry, and sick, ultimately confuted the current theories of social Darwinism.[69] If, in Fourier, excellence, novelty, and digestibility are the "healthy" results of a ludic "hygienic war," applied to the alimentary field, degeneracy, fatigue, and sickness are the actual risks of the hygienic practices Marinetti ideally linked to war in his 1915 text. His desired "alchemical" transformation of the "heart" into a "stomach of the brain" is thus undermined by a war that actually "destroys cellular chemism [...] hurts the reciprocal and delicate interplay between the bowels and the brain."[70]

However, as Rouvillois observed, a "hygienic compulsion" is inherent in any utopia (in Schaer, 319). The idea of ultimately managing to "break with [man's] bestial origins" – that is, in these circumstances, Marinetti's "ants," "mice," "cats," "oxen," or Fourier's less specific "unrefined gluttons" – may occur either because of a peaceful, selective "gastronomic fight" aimed at achieving the tastier and most digestible food (Fourier), or because of a war that holds the (false) promise of regeneration and, with it, of a healthier, "lighter" and more refined race (Marinetti). Although Marinetti and Fourier clearly did not envision similar scenarios when they invoked the notions of "harmony" and "war," one may still safely observe that at a certain point, in both writers, the issue of selection, be it of food or men, and the path towards the achievement of a "better" state simultaneously involve and feed some sort of conflict. Thus, I end my comparative glimpse into the utopian alimentary discourses of Fourier and Marinetti with the suggestion that another will have to decide if the latter dynamics are a symptom of the prophetic or lunatic quality of their work.[71]

69 See Pogliano, 64.

70 Pogliano, 64.

71 Larizza notes that "there are many interpreters who insist on Fourier's folly" (Larizza, 3n).

1.3 An Appetite for Matter: Sex and the Inorganic in F.T. Marinetti's Erotic Short Stories

> La morte forse non è che un punto dell'infinito dove si comincia a gustare la vita.
>
> — F.T. MARINETTI

> Things do not exist without being full of people.
>
> — BRUNO LATOUR

My previous discussion of "A Meal that Prevented a Suicide," the short narration at the beginning of *The Futurist Cookbook*, allowed a preliminary taste of how its author(s) (and, by extension, futurism itself) envisioned the relationship between food, sexuality and femininity. It is, however, through a consideration of the substantial contribution of Italian futurism to the literary trend of the "social-erotic" novel – a popular genre of consumer fiction widely published in the early decades of the twentieth century – that such a relationship can be better analysed and assessed.

So far, few scholars have addressed the subject and not many publishers have been interested in reprinting anything but those texts which Marinetti himself authored or co-authored. As a consequence, while his *Novelle colle labbra tinte* (1930) (from now on: *Novelle*) and his earlier collection of short stories *Scatole d'amore in conserva* (1927), his "didactic, 'hygienic' treatise" (Sartini Blum 90) *Come si seducono le donne* (1917), his novels *Isola dei baci* (with Bruno Corra) (1918) and *L'alcòva d'acciaio* (1921) have enjoyed recent re-publications, other representative examples of this futurist narrative sub-genre – for example, Emilio Settimelli's *Donna allo spiedo* (1921), Francesco Cangiullo's *Ninì Champagne. Romanzo vivo* (1920), Fillìa's *L'uomo senza sesso* (1927) and *Lussuria radioelettrica* (1925) or Angelo Rognoni's *Carne* (1918) – wait to be rediscovered.

Regardless of their critical or publishing status, the revealing titles above confirm the general tendency of futurism to imagine woman as a sexual object to be seduced and possessed. Moreover, romantic love is viewed as a quickly disappearing sensation, to be substituted with one, which, like food, can be possibly put "in conserva." Finally, sexuality is considered a function that oscillates between the spheres of the organic and the inorganic – or, to follow the suggestions of the titles, between "carne" and "champagne" on the one hand, and "acciaio," "spiedo," and "radio-electricity," on the other.

Not surprisingly, Cinzia Sartini Blum, one of the scholars who have written influential pages on the social-erotic genre as an important part of the futurist discourse on love and war,[72] observed that "the most revealing emblem of Marinetti's attitude toward reality is the recurrence of alimentary imagery." (With some distinctions, this statement also applies to many of Marinetti's companions).[73] She thus suggests that the "erotic scenarios" provided by these narratives, in which "gustatory and alimentary metaphors are stock material" (93–4) are a privileged context for examining the close link among sex, food, identity, and aesthetic creativity.

Sartini Blum's analysis, in this particular section of her study, makes use of feminist and psychoanalytical theory and the methodological lens of gender in order to demonstrate how Marinetti's and futurism's aggressive, assimilative, carnivorous rhetoric is a mechanism which, simultaneously betrays an anxiety and reacts to the threat that "the engulfing feminine"

72 See also the methodologically similar works of Lucia Re and Barbara Spackman, just to mention two of the scholars who, in their analysis of the feminine figure in various contexts of the futurist avant-garde, have observed women's objectification/reification and, in general, regarded it as among the first steps of the "virilization" and subordination of women to the male universe. As will become apparent in the rest of this section I will be lingering a little longer on some additional, potential implications of this objectification process.

73 A collection of trite "social-erotic" short stories like Settimelli's *Donna allo spiedo* (1921), for example, does not seem to share any of the aesthetic/stylistic concerns or the ironic self-awareness with Marinetti's. Although Settimelli's tales are certainly interesting from a historical and sociological perspective and for the usual recurrence of sexualized alimentary imagery, they develop only the most superficial aspects of futurism's poetic. As I wish to ultimately argue, Marinetti's production may instead combine those same stereotypical features of the genre necessary to lure the "popular," average reader, with some deeper intuitions.

(82) represents for male identity and power. As she puts it: "the increasingly active role played by women in modern society was a persistent source of concern for the futurists" and, therefore, "a healthy, ego-boosting 'diet' of seduced women [was deemed necessary] to sustain [contemporary man's] masculine identity" (81, 89).

Her observations cover several contexts and times in which Marinetti addressed "questions of sexual politics" (81) but, as far as the specifics of the erotic genre are concerned, focus mostly on Bruno Corra and Emilio Settimelli's preface to Marinetti's *Come si seducono le donne* and, with regard to his *Novelle*, on "La carne congelata" (a.k.a. "Come si nutriva l'ardito"). Sartini Blum convincingly illustrates both the connection between ingestion, assimilation, identity, and the cycle of artistic production and consumption on the one hand, and the "ambivalent effects of desire" which, playing out through opposite dynamics of objectification and idealization, end up "dissolving the boundaries of [male] identity" (97).

Building on Blum's line of interpretation, my objective here is to examine a wider sample of Marinetti's short-stories and their sexualized alimentary imagery in order to suggest an additional, though more marginal, perspective on his "social-erotic" production – one that relies on some of Mario Perniola's stimulating reflections in his *Il sex-appeal dell'inorganico*, and is less dependent on the gender factor and more attuned to some of the fundamental principles and positions of futurist poetics.[74]

Let me begin by briefly recalling a couple of these principles, beginning with Marinetti's notorious "war against time" described in the founding manifesto (1909).[75] First, his Nietzschean rejection of history is aimed at unmooring futurism from any past literary tradition in search of a pure origin and, thus, at avoiding the stifling "sickness" that would derive from its insertion into the historical process. Second, there is his

74 Mario Perniola, *Il sex-appeal dell'inorganico* (Torino: Einaudi, 1994).

75 I am obviously referring to the eighth point of the first manifesto of Futurism: "Noi siamo sul promontorio estremo dei secoli! ... Perché dovremmo guardarci alle spalle, se vogliamo sfondare le misteriose porte dell'Impossibile? Il Tempo e lo Spazio morirono ieri. Noi viviamo già nell'assoluto, poiché abbiamo già creata l'eterna velocità onnipresente" (Marinetti, *TIF* 11).

related position with regard to the question of desire: his concern with the possibility of being able to renew and refresh indefinitely one's desire and pleasure. As noted earlier, a specifically alimentary text such as *The Futurist Cookbook* (1932), in which desire (and/or appetite) must be kept alive by emphasizing variety, by avoiding boring companionship at the table ("abolizione dell'eloquenza e della politica a tavola" [33]), and by looking for unique combinations of unusual ingredients ("abolizione delle tradizionali miscele per l'esperimento di tutte le nuove miscele apparentemente assurde" [29]) well exemplifies this preoccupation. As Onfray summarized in regard to *The Futurist Cookbook*, "L'important est de produire le désir" (Onfray 164). I now wish to consider Marinetti's social-erotic short fiction and its representation of sexuality in the light of the two principles mentioned above.

It will probably be one more sign of the many contradictions and ambivalences that characterize "Marinetti's discourse on love" (Sartini Blum 81), but I find it slightly problematic to reconcile the virile, "carnivorous" aggressiveness widely displayed in his texts – one which, for the good reasons mentioned by Sartini Blum, almost univocally aims at metaphorically ingesting and sexually possessing as many women as possible – with the fact that such a libidinous, "orgasmolatric" behavior (a major purpose of which is the temporarily satisfaction of the male body's sexual hunger through the consumption of a series of females and the achievement of an ephemeral instant of pleasure, simultaneously) has the effect of subjecting that same body to the flux of time and unavoidably exposing it to the malady of history.

It is well known that both sex and food have something to do with the starving, needy nature of the human condition, with its physiological finitude and perishability. Having sex and eating are activities that could be viewed as (usually) pleasurable and necessary relieving distractions, which, while they make human beings temporarily forget the precariousness of their life, unavoidably remind them of their existence engraved in history. As Bettina Knapp reminds us when she discusses "The Sex Machine" in Alfred Jarry's fantastic novel *The Supermale* (1900) "to permit emission of semen is to fall into the law of time and death: that is, to be the victim of empirical existence. Under such conditions, lovers would be merely libertines and not the detached, Godlike beings they seek to become."[76]

In other words, Marinetti's original, utopian attempt to subtract futurism from history[77] and his effort to exorcise as much as possible the ghost of corruptibility (just remember the character of Saint Putrefaction in *Le Roi Bombance*) would seem to render problematic the male characters' drive to seduce, metaphorically ingest, and possess one woman after another, independently of their age, race, or social condition,[78] in order to reach a fleeting climax of sexual-alimentary orgasm or, else, a "ventre rimpinzato di piacere" (Marinetti, "La domatrice di leoni" in *Novelle* [185]). Futurism's technological infatuation and its subsequent desire to progressively free the human body from any physiological need (be it alimentary, as the search for "leggerezza" in *The Futurist Cookbook* suggests, or sexual, as implied by the sexless creation of the "Godlike being" Gazurmah in *Mafarka*)[79] further emphasize this point.

Although this kind of aggressive, orgasm-oriented sexuality from a feminist angle can constitute a defensive reaction against and response to the threat posed by the abject feminine to male identity, it also exposes the (male subject) owner of such identity to another, perhaps even more dangerous threat. An insatiable appetite for flesh, rather than protecting the futurist male identity and/or contributing to projecting the futurist subject

76 Bettina Knapp, "Jarry's The Supermale: The Sex Machine, the Food Machine, and the Bicycle Race. Is it a Question of Adaptation?" *Nineteenth-century French Studies* 18.3–4 (1990), 492–507, (505).

77 Or, as Somigli writes, to "generate a textual experience which [...] transcends time and space itself, and reaches towards the [Godlike] 'absolute' of the poetic page." Luca Somigli, "On the Threshold: Space and Modernity in Marinetti's Early Manifestoes and *tavole parolibere*." *Rivista di Studi Italiani* 17, n. 1 (1999), 250–73, (259).

78 Consider Fatma, the Arab prostitute and "femmina qualunque" in "Cacce arabe" and, at the opposite end of the social spectrum, the "Marchesa Marcella."

79 An explicit title such as Fillia's *Uomo senza sesso* (1927) would seem to confirm this tendency even more clearly. On this trend towards a bodily "leggerezza," see also Karen Pinkus's acute observations on 1930s advertising and the "phenomenon of bodily disappearance" in her *Bodily Regimes. Italian Advertising under Fascism* (Minneapolis and London: University of Minnesota Press, 1995), 196.

towards the realization of his theorized "metallized," a-historical, absolute existence free from the risk of abjection and the fetters of time and death, contains *in nuce* the cause of his own dissolution and self-annihilation. At the same time, however, this dissolution may paradoxically constitute one of the necessary premises with which to begin a process of timeless renewal.

One could argue that the *Novelle* themselves, as a "cultural artifact which inevitably exist[s] upon the historical plane" (Somigli 1999, 271), render any attempt to evade history futile, especially if one imagines the ink projected onto these pages as symbolic of the fictional semen that is, more or less explicitly, projected within the various stories themselves.[80] However, my objective here is precisely to underline further the apparent contradictions indicated above and, in particular, to examine the rhetorical strategies and aesthetic solutions (Somigli aptly used the expression "balancing act" [271]) which Marinetti, adopted time after time to cope with the paradoxes of a situation he himself created. (I am not discussing here Marinetti's degree of consciousness or intentionality, just what seems to happen in his texts.)

An initial example of such rhetoric may perhaps be found in Marinetti's "La simultaneità in letteratura" (1930), which constitutes the preface to the Vallecchi edition of the *Novelle*. Here Marinetti significantly writes: "Le vecchie forme del romanzo e della novella sono sciupate e inadatte a divertire, *nutrire e dinamizzare gli spiriti*" (22, italics mine). What I find worth noting is that an expression such as "nutrire e dinamizzare gli spiriti" does not sound particularly appropriate since he is introducing a series of erotic tales where, as repeatedly observed, that which is primarily "fed" (with a diet of seduced female flesh) is not so much the spirit but, rather, the male body. The focus, however, reading beyond the stock-material conventional quality of alimentary metaphors in futurist erotic literature, is here shifted from the corporeal to the spiritual. Why does this initial

80 For a discussion of futurism's "materiality of language" and the "avant-garde [...] general engagement with the physicality of concrete and textual matter" see (Delville 127), and Fausto Curi's "Una stilistica della materia" in *Tra mimesi e metafora. Studi su Marinetti e il futurismo* (Bologna: Pendragon, 1995), 47–79.

nutritional reference regard a spiritual and aesthetic sphere rather than merely an empirical, physical one, as the latter would seem more obvious and appropriate given the popular, target audience and the light, prevalently consumeristic nature of these texts? Why couldn't he simply write "nutrire e dinamizzare i corpi," since most of his erotic short stories apparently deal with that?

My guess is that Marinetti's awareness of and frustration with the limitations and temporality of the body and his utopian hope (in a sense, similar to the one alluded to in *Le Roi Bombance*) that a well-fed spirit may, in turn, sustain the artist's search for the "ideal" and the Absolute, still haunted him and thus sporadically emerged even in these later writings. If we take Marinetti's words literally, his "hunger for the Absolute," although disguised under a different, more "pragmatic" shape (Sartini Blum 89), may still be partially present. In theory, then, his *Novelle* are intended to "feed and dynamize" a genderless, eternal spirit, and not exclusively the virility and vitality of the decaying, post-war male consumer body.[81]

Of course, Marinetti's programmatic call to "diminuire il bisogno d'affetto" and to reduce romantic love to the "conservazione della specie [...] semplice funzione corporale, come il bere e il mangiare" (*TIF* 300–1) which, arguably, his *Novelle* intend to exemplify artistically, makes it questionable to suggest any privilege of the spiritual at the expense of the corporeal in the texts I am dealing with. However, what I especially wish to emphasize at this point is that when Marinetti chooses the word "spirits," with its time-defeating connotations, his obsession with time re-emerges along with his attempt to circumvent it artificially at the textual level.

It does not seem by chance that after reducing love to bodily functions in his 1915 "L'uomo moltiplicato e il Regno della macchina," Marinetti adds that "*L'uomo moltiplicato* che noi sogniamo, non conoscerà la tragedia della

81 As a matter of fact, Blum herself acknowledges in Marinetti's discourse on love the presence of "occasional residues of nostalgia for the lost 'mystery' and 'absolute value' of *amore*" (Blum 89). On the ambivalences of modernity, see also Somigli's *Legitimizing the Artist. Manifesto Writing and European Modernism, 1885–1915*. (Toronto: Toronto University Press, 2003).

vecchiaia" (*TIF* 301: italics in original), thus subliminally linking love and desire with a dream of eternal renewal, youth, and vigor.

The paradox seems to be that, even though Marinetti writes about eventually "killing love" for his ideal of "uomo futuro" (*TIF* 300), as soon as he reduces romantic love to drinking and eating ("il bere e il mangiare"), he suggests both that love is still a physiological necessity for man and that, although under renewed, artificial shapes (e.g., "il cuore deve diventare [...] una specie di stomaco del cervello"), it will need both to be preserved and to characterize "i giovani maschi contemporanei." The latter, in fact, will be entertaining "*infinitamente* il loro sesso con contatti femminili rapidi e disinvolti" (*TIF* 301: italics mine).

The title of Marinetti's first collection of erotic short stories, *Scatole d'amore in conserva,* is particularly revealing in this sense and may help to further clarify my assertions. A canned ("in conserva") love, arguably with an indefinite expiration date, is the one that also needs to "conservare [la specie]" (*TIF* 301) or preserve the human species. At issue here is certainly the superfluousness and "total rejection of love" (Sartini Blum 89) characterizing the mechanized, ideal superhuman futurist type, in accordance with the ("anorexic") reduction of human physiological needs mentioned before, but also the creative attempt to cope with the realization that love should be saved (or "conservato") from the ravages of time precisely in order to assure the existence of the "uomo futuro."[82]

If, in addition to the threat represented by the feminine to the integrity of male identity discussed by Sartini Blum, other, apparently unsolvable preoccupations (for example, the risk of falling into the "law of time and death"; the search for a kind of love and desire which could also be an "infinite distraction" rather that a mere instant of bliss, and, as a consequence, for a kind of utopian, absolute sexual-alimentary-aesthetic experience which goes beyond the body's perishability) may be tangentially at play in Marinetti's social-erotic production, it is time to take a closer look at some of his short stories and discuss how these impasses may have been confronted.

82 On the "formal and ideological implications of food-packaging," see Delville 64.

The short story "Miss Radio," in *Novelle*, may be particularly appropriate in this regard. Its plot revolves around a transatlantic trip to Buenos Aires that Aldo Vetri, the young "inventore d'un nuovo motore," undertakes "chiamato dal Governo argentino [...] più per dimenticare una donna che per realizzare la sua fortuna" (61). Aldo, who travels with his older brother Luciano on the ship *Luce*, has been betrayed by his former lover, Millina, "la tipica sensuale spudorata con relativa vernice di purezza," nicknamed "Miss Radio" because she used to tell him that "se la vita ci separa, le care nostre onde elettriche ci congiungeranno per sempre!" (64). During his journey Aldo, still obsessed with the woman, repeatedly feels her presence around him and imagines sensing her presence in a variety of things. First, it's the ship's turn: "Il nostro transatlantico [...] freme e palpita come lei! Si! Come lei! Luciano credimi, non è un modo di dire. Il nostro transatlantico è realmente una donna! [...] La nave [...] precisava la sua femminilità dando i fianchi sempre più veloci al tattilismo frusciante della seta carne marina" (62–3); then he sees her in a storeroom fridge, as a "ghiotto and saporito" statuesque ghost made of "ghiaccio e sale," ready to be "mangiato di baci"; then in the burning "levigatezza" of the ship's engine and, finally, as her talking profile appears in a little cloud of "fumo perlaceo agile e grazioso": "Mi fuggi, Aldo, e mi troverai dovunque! Anche in cielo! Corro sulle onde della radio!" (68). Millina, in a technological, futurist version of every classical water metamorphosis, has ultimately transformed into something even more eternal than water-waves: a radio-wave, pure electric energy.

Leaving aside the fantastic aspect of the story and the originality of Marinetti's invention, I wish to now reconsider Millina's predictable process of reification and objectification, while emphasizing both the affinity that such a process seems to have with that "internalization of technological apparatuses" (Brown 12) initially mentioned by Walter Benjamin in his reflections on people and things in a modernist context and, of course, its recurring presence within these social-erotic short stories.

If considered in terms of sexual appetite, as would seem appropriate given the context, the fact that Millina's original, organic body is represented and perceived as transformed into a series of inorganic and yet eventually sentient, live things, while Aldo's body itself is more and more reduced to a bundle of "multiplied" sensations and desires for his lost lover, may suggest that a peculiar kind of sexuality can be traced here. This sexuality partially recalls the one that the philosopher Mario Perniola describes in his *Il sex appeal dell'inorganico* as the "sessualità neutra, sospesa e artificiale della cosa che sente," one that "emancipa la sessualità dalla natura e l'affida all'artificio" (5).[83]

The very title *Novelle colle labbra tinte*, by the way, may reveal an "inorganic" tendency towards "treating the word as a sentient thing,"[84] since it achieves what Perniola describes as the simultaneous sexualization of the literary word (the "Novelle" possess "labbra" themselves), and the subsequent desexualization of the body or a specifically sexual organ (the "labbra" are associated with a book, with the written word). In other words, here the book paradoxically goes in the direction of the organic while the body, the feminine body in particular, does the opposite.[85]

This said, the utopia of "freeing sexuality from nature" (and both the tendencies towards reification and the technological present in futurist social-erotica seem to be steps in this direction), means suspending it from the irreversible flow of time, and being able to experience it indefinitely, possibly with the same intensity. In simple terms, if I could manage to leave

83 Perniola's debt to Benjamin's reflections on the relationship between human beings and the world of things is clearly stated at the beginning of *Il sex appeal dell'inorganico*. On a different note: futurism's understanding of erotic experience as objectification and reification of the woman's body seems to be at the exact antipodes of Georges Bataille's position on the subject. According to Direk, for Bataille, in fact, "erotic experience aims at destroying the thing in the body. As an inner experience, it aims at the liberation of the body from being a thing" (111). Zeynep Dyrek, "Erotic Experience and Sexual Difference in Bataille" in *Reading Bataille Now*, ed. Shannon Winnubst (Bloomington: Indiana University Press, 2007), 94–115.

84 "Una *philosophia sexualis* è possible solo dal momento in cui la parola è trattata come una cosa che sente" (Perniola 1994, 157).

85 "Più in generale questo primo modo di sentire il corpo come cosa non è privo di legami con una desessualizzazione degli organi sessuali, e reciprocamente con una sessualizzazione degli altri organi" (Perniola, *Il sex appeal*, 41). Marinetti's noted reduction of the heart to a sort of "stomaco del cervello [...] perché lo spirito possa entrare in azione" (*TIF* 300) comes to mind again at this regard. As the heart loses its centrality in love (and sex), the spirit allusively "rises up" and acts.

my organic nature and yet somehow still be able to feel and desire, I could also cheat time and prolong my sensations indefinitely.[86]

As Perniola observes when he examines some of the features of "estetica neutra" in which "trova spazio e riconoscimento tutta la sessualità cosiddetta perversa [...] sadismo, masochismo, feticismo e necrofilia," the "trasformazione della carne in cosa, dell'organico in inorganico [...] desta un'eccitazione che è accresciuta dal pensiero di estendere la sessualità al di là della vita e al di là di quella vita purulenta e verminosa che la decomposizione cadaverica continua nella morte" (Perniola 1994, 102).[87]

Returning to "Miss Radio," the literally ethereal and artificial words that Millina, now reduced to an electric frequency, addresses to Aldo at the conclusion of the story ("mi troverai dovunque! Anche in cielo!" [68]) and that six months later mysteriously come out of the mouth of the heroic radio officer while the *Luce* sinks ("Resto al mio posto qui dove sto bene e dove fra poco Lui [Aldo] mi raggiungerà" [68]) do not seem to exclude an allusion to the possibility of a "non-natural" sexuality and an infinitely renewable, constant state of excitement and desire, both traceable "al di là della vita" (Perniola 1994, 99) (i.e.,"in cielo").

Perniola's additional reflections on a "sessualità neutra" that "nasce [...] dalla approvazione incondizionata dello spazio illimitato aperto dalla scomparsa del soggetto" (56); his unorthodox reading of sexual perversions and fetishism; his insightful comments on "bodies as clothes," or his individuation of the primary experience associated with inorganic sexuality in a "dissolvere le forme, far valere i principî della continuità e della transitività, passare da elemento a elemento senza iati o fratture, far saltare il primato

86 On the link between subjects, objects and temporality, Bill Brown observes that "W.J.T. Mitchell makes it clear that the discovery of a new kind of object in the eighteenth century, the fossil, enabled romanticism to recognize and to refigure its relation to the mortal limits of the natural world" (qtd. in Bill Brown, "Thing Theory." *Things*. Ed. Bill Brown. Special issue of *Critical Inquiry* 28.1, 2001, 1–22, [16]). These "limits" are precisely those that the modernist avant-garde is trying to redefine.

87 Perniola also interestingly notes: "La filosofia della cosa libera la sessualità dalla dipendenza dall'organico, in cui tanto la psicoanalisi quanto il femminismo l'hanno mantenuta" (Perniola 1994, 60).

della sessualità genitale e organica, sessualizzare il mondo" (159) acquire even further relevance and a peculiarly familiar aura when applied retrospectively to the futurist poetics and, in particular, to some of Marinetti's other social-erotic short-stories.

For example, "11 baci a Rosa di Belgrado," the epistolary short story which opens the *Novelle*, seems to confirm the predictable, objectified role of women as tasty morsels to possess/eat/drink: "golosamente vi mangio la bocca"; "suggo giù il miele della vittoria"; "spremuta di baci" (33, 44, 47). Or it may represent a move towards a technological desexualization of the organic (the prize for the kissing competition is a "strambo gioiello in forma di labbra meccanizzate" [29]) and the sexualization of the inorganic ("novelle palpitanti;" "i silicati virili della sabbia;" "un piccolo agile intimo aeroplano quasi corporale" [50, 52, 55]). But it may also be characterized by some of the other features which Perniola associates with a "sessualità inorganica."

In this sense, one could perhaps talk both of "subject disappearance" as well as "dissolution of forms" when, in his second letter, "Il bacio ginnasta," the male narrator/protagonist expresses to Rosa the will that his life "vuole mutarsi in farina per il vostro pane quotidiano" (36). Besides indicating that in these erotic stories women are not always the only available sexual "edible matter" (men can be objectified and eaten too),[88] the protagonist's declaration may suggest that by transforming himself into an element, even though one still organic, such as "farina," with the hope that he will be consumed by Rosa day after day ("quotidiano"), he is advocating an essentially virtual sexuality, one which, if we may again appropriate and adapt Perniola's words, aims at being "non [...] precaria ed effimera come quella naturale, ma sempre disponibile nella sua vertiginosa artificialità" (*Il sex appeal* 39). And, of course, the reference to the transformation of "life" into "flour" and then into "daily bread" contains also a not-so-subtle religious allusion to the miracle of transubstantiation, which, consistent with my argument, further alludes to notions of timelessness and eternity.

88 In this regard, see also "Consigli ad una signora scettica" and "La guancia" (both in *Novelle*).

A similar situation, in which the lovers' sexuality almost pantheistically expands into and assimilates the surrounding environment, arguably seeking to achieve a permanent continuity of sensations, emerges first in the sixth letter ("Il bacio nuotato"), where the two merge and dissolve into the sea, losing their identities and becoming one with the water:

> Le vostre ghiandole sublinguari esprimono già una saliva piena di ptialina deliziosa. Assaporo nella schiuma i cloruri di sodio potassio e magnesio. Facciamo parte del potente complesso marino. [...] Baciarci a bocca larga bevendo tutte le salive la tua la mia e quella di Venere, il mare. Tutto. (46)

It occurs again in the ninth letter ("Il bacio piovoso"), in which the union between the two lovers is compared to a basically inorganic "amplesso sensuale tra la pioggia e i silicati virili della sabbia" (52). Finally, in "Bacio tropicale," when the protagonist yearns to "rimpinzare [Rosa's] [...] vene di piacere [...] Senza l'assurdo intoppo o nodo della morte." (48), it is hard not to be reminded once again of the attraction and appeal that a deathless pleasure or an orgasm-free sexuality – which can artificially go beyond decomposition and death – has for Marinetti and his characters (cf. Perniola 1994, 99 and 102).[89]

It is, however, precisely a death by suicide that concludes "La logica di Ahmed Bey," a short story which, at first glance, would seem both to undermine what I noted above and to belong more to the war genre than the social-erotic. So, one may legitimately wonder, how can Ahmed's, not to mention my "logic," be explained here?

At the end of the journey, the transatlantic ship "Massilia" is approaching Rio de Janeiro "come una sposa distratta" and, as such, performs a sort of Sapphic sexual act with the "virgin," feminized, edible landscape around it:

> Gioia di entrare nella tenerezza carnale della Baia [...] La nave ha alzato tutte le sue metalliche orecchie bocche tasche da vento per udire bere insaccare golosamente note d'oro liquido, fluidi pani dorati, vermigli zuccheri filati, languide frecce di miele (107–8).

89 There is probably no need to recall the definition of orgasm as "little death."

The Turk Ahmed, whom the narrator/fellow traveler had first met fifteen years earlier during the Balkan War, whispers the words "C'est trop beau! C'est plus beau que le Bosphore! Pauvre Stambul!" and then "balzò sul parapetto, e [...] precipitò nelle onde" (*Novelle* 108–9).

As we are eventually told, Ahmed's suicide has a "causa [...] estetica e patriottica" (109). Apparently, the overwhelming beauty of the Brazilian landscape has killed him. If, while he was alive, he was not able to interiorize and possess such an aesthetic superiority, as soon as the combination of "Bosforo di porpora" and Brazilian ocean wraps "morbidamente" around his dead body, he appears "finalmente liberato dal tempo e dallo spazio. Pacificato" (109). The beauty of the "Baia," which previously could be metaphorically penetrated, eaten, and possessed only in an ephemeral way, may now be possessed for eternity. As Ahmed lets himself dissolve into nature, he becomes one with the seascape that surrounds him and, by moving towards an inorganic state, he evades the limitations of desire and the frustrations of temporality. In his condition of a "thing-like" corpse ("cadavere") who nonetheless may still mysteriously be able to embrace, "feel" and "taste" the world around him/it, Ahmed perfectly incarnates the concept expressed in the quotation at the beginning of this essay and reasserted in the following one, also from "La locomotiva blu": "Evidentemente gusteremo la vita quando saremo morti" (256).

The gesture of "farsi mondo," which tends to "abolire la distanza che separa l'uomo dalla cosa" (Perniola, *Il sex appeal* 123), together with the programmatic futurist "multiplication of the I," surfaces in an even clearer fashion in "Fabbricazione di una sirena." In this story, the panic urge to assimilate the surrounding reality – "Languidamente pesco colgo e raduno tutti i colori sapori odori rumori del mare" (114) – goes hand in hand with the expressed wish to lose one's humanity in order to be transformed into a multiplied series of progressively more ethereal things: "Mi compiaccio di perdere la mia umanità. Pesce. Diamante. Raggio. Bolla d'aria. Vibrazione. Sfumatura" (*Novelle* 114). As the title makes clear, the final objective of this marine embrace is to build a mermaid piece by piece by poetically assembling the various elements offered by the sea: "spirale d'acqua quasi coscia [...] Due listine di alga rossa: labbra! [...] la sirena sorgiva d'acqua cristallo carne oro perle burle sorrisi carezze fughe ritorni avorio alabastro nickelio coralli barbagli ventagli di schiuma che fuma" (115). In what may perhaps recall the parthenogenetic, woman-free birth (and subsequent flight) of Gazurmah, this completed creature who lives solely "per virtù di un poeta" (117) is artificially created by the combination of a variety of materials only to be ultimately kidnapped and hauled up into the sky by the protagonist on the "idrovolante del Capitano Keller" (117).[90]

The word has artificially become flesh (or perhaps more appropriately given its assembled, elemental origin, a "thing that feels") for the virtual pleasure and, arguably, the "eccitazione costante" (Perniola 1994, 39) of its creator: "Sono nuda, ma s'affollano intorno a me le più belle vesti e potrò vestirmi riccamente per i tuoi inverni" (*Novelle* 116). This statement, incidentally, cannot avoid suggesting the potential subordination, in terms of sexual attraction, of the "sentient," naked body of the mermaid to the even more clearly inorganic things (i.e., the "belle vesti") which cover it.

"Fabbricazione di una sirena" thus alludes to the sexual pre-eminence of clothes over the body who wears them, through which it would be possible to inorganically "sentire il corpo proprio o altrui come un vestito" (Perniola 1994, 58) and evade a past, exclusively organic-centered way of understanding sexuality. However, this allusion seems, nonetheless, to be immediately contradicted by two other intriguing *novelle*, "Fa troppo caldo" and "La guancia," in which the body's bare nakedness, with all its implications and potential complications for my argument, apparently takes its revenge.[91]

"Fa troppo caldo" begins by displaying the well-known, aggressive, carnivorous behavior of the stereotypical futurist male engaged in educating his Russian lover during a day at the beach: "Io odio i vestiti! Mi piaci in costume da bagno. Meglio senza. [...] Mi piaci nuda!" (192). Its

90 In line with the general trend of my observations, since a "sirena" does not fully belong to the divine, animal, or human sphere, she could quite easily fit the category of "cosa che sente" (see Perniola's chapter on "Dio, animale, cosa").

91 And yet, in his "Prova d'una veste" in *Il poema non umano dei tecnicismi* (1940), Marinetti would write: "Anzitutto finiamola colla vantata superiorità dell'umano [...] Stoffe stoffe [...] e poiché sono siete tutte sensibili pensanti parlino parlate parla finalmente" (*TIF* 1190).

even longer "Variante" then humorously reasserts and expands upon both libidinous virile behavior and the apologia of nakedness with the description of a Dionysiac, orgiastic scene set in the "postribolo più frequentato," that culminates in a provocative, *absurdiste* manifesto "against clothing":

> Le donne a due a due arenate sui divani fioriti di viole come le spiagge di Mitilene. I maschi in agguato come sottomarini. [...] Ci saranno molti letti con miscele, acrobatismi, dinamismi e compenetrazioni di donne [...] Viva la nudità! Abbasso i vestiti (194).

While it would be ridiculous, in this context, to attempt to deny the presence of a debasing, consumptive, orgasm- and genital-centered sexual attitude against women, or the declared primacy of the naked body over the clothed one (we are in a brothel, after all!), the "Variante" of "Fa troppo caldo" may also lend itself to some additional – perhaps less obvious – reflections.

First, as a marginal note, one should probably mention Marinetti's mock-ironic and demystifying undertone in these pages. Besides implying a surprising self-awareness, this tone is not only directed towards himself as the founder of futurism ("le guerre saranno completamente abolite il giorno in cui l'umanità sarà tutta nuda" [194]), but also targets and caricatures no less than the quintessential figure of virile sexuality and political bellicosity of that period:

> Poi ripeterete questi argomenti perentori, parlando nudo al balcone ai sottostanti cittadini [...] quando [...] nasceranno sui crani calvi dei dominatori il nuovo naso, gli occhi nuovi e la bocca nuova, avremo finalmente l'età paradisiaca della pace assoluta (195–6).[92]

Second, and more important to the general objective of my analysis, the same lexicon in an expression such as "gli svariati inscatolamenti, le

92 I am obviously referring to Mussolini, who, at the time, perfectly fit the combined image resulting from putting together "verbose balcony speeches" plus "dominating baldness." In this sense, see also the related allusion to the "rubinetto fresco della vostra eloquenza" (195).

saldature, gli innesti, le trecce, i nodi dei corpi oliati di sudore" (194) unavoidably creates the mental picture of a kind of sexual intercourse that lies at the liminal border between the organic and the inorganic, one in which the "performers'" own identities and subjectivities disappear and, thus, could perhaps be legitimately defined as "quasi-things that feel," as well as "quasi-bodies that sweat."[93]

This same image of an entwined, mechanical, and almost indistinguishable, "neutral" bundle of interpenetrating bodies, somehow fixed in their tableau-like appearance for the sake of the viewer, suggests, in turn, a comparison with Perniola's notion of "campo di carne." With this expression, or the equivalent one of "corpo senza organi," he individuates one of the modalities of the "cyborg filosofico-sessuale" and the sexually neutral, timeless "punto in cui [la carne] raggiunge la fissità imperitura delle cose" (Perniola 1994, 42–3). That is to say, Marinetti's "miscele" and "compenetrazioni" may partially recall and anticipate that situation of undifferentiated sexual anonymity and impersonality which the philosopher characterizes as "neutral," "thing-like" and "sempre disponibile."

All the hygienic prescriptions that appear in the final manifesto, such as 4: "curare il proprio corpo," 6: cleanse humanity from "ogni sozzura," or 8: the total destruction of "tutti i peli dell'Umanità," beyond their shock-value, contribute to reinforce the evocation of an unnaturally naked cyborg-like figure. Or, if we push this analysis a final step further, someone whose artificially smooth and hairless skin constitutes also his only sexualized "vestito" and who, without the need either to "correre verso l'orgasmo" (Perniola 1994, 43) or, similarly, to have any "ossessione lussuriosa," will instead have access to the "era paradisiaca della pace assoluta" (*Novelle* 195–6) and to a virtual, death and time-defying experience of sexuality.

93 Some of Brown's observations come to mind and seem particularly relevant at this point, especially when he notes that "Bruno Latour has argued that modernity artificially made an ontological distinction between inanimate objects and human subjects, whereas in fact the world is full of 'quasi-objects' and 'quasi-subjects,' terms he borrows from Michel Serres. Benjamin makes it clear that the avant-garde worked to make that fact known: modernism's resistance to modernity is its effort to deny the distinction between subjects and objects, people and things." (Brown, 12).

The simultaneous presence of clothing metaphors, the search for an "absolute" nakedness, with the added background of a vampiric, extreme, and totalizing kind of sexual experience, characterizes also "La guancia" which, according to Domenico Cammarota, editor of the *Novelle*, is "il più bel racconto dell'antologia, dove il vampirismo sessuale della protagonista anticipa di settant'anni la poetica horror della 'Nuova carne'" (*Novelle* 13).

The narrator begins by directly addressing the "cara biondina" who is the melancholic protagonist of the story and, through a literal dissection of the apparently enviable relationship that she has with her lover ("un uomo intelligente volitivo [...] non ama che te [...] ti rimane fedele" [217]), proceeds to explain to her, and the reader, the mysterious origin of and the relative solution to her "melanconia" and lack of satisfaction. The explanation that he gives her essentially regards the fact that her lover, after every intercourse, "rimane [...] sempre lui. Cioè padrone di sé [...] Il sangue ricco che gli scorre sotto la pelle [...] torna a ricolorargli le gote quando i tuoi baci [...] lo hanno fatto godere [...] fino al pallore lividazzurro della morte" (*Novelle* 217–18). From the perspective of her sexual gratification, the woman's problem seems to be that her man is still alive and well after their sexual embrace; that he recovers from the ephemeral, death-like experience of the orgasm, and, finally, that his gender and identity are still unvaried: he is still a "him" who is "padrone di sé" (217).

This "mistero" receives further clarification when the narrator, in his role of sexual counselor, also tells the "biondina": "Tu [...] soffri di sentirlo ancor vivo staccarsi da te, baciarti con grazia e balzar giù dal letto con la sempre eguale elasticità delle sue gambe giovani. Questa angoscia turba la tua felicità" (218). Once again, his remark reaffirms that the crucial cause of the woman's anguish and the obstacles to her full pleasure and gratification are the fleeting, short-lived quality of the experience and the vitalist, energetic humanity of her lover. The fact that he behaves and reacts like a normal and eventually predictable and dull, "organic" human being, rather than as someone or some-*thing* else, seems to be the problem here.

As the narrator individuates the nefarious effects that traditional sexuality has on the woman, not only does he implicitly recognize its inherent lacks and insufficiencies and the need for a more comprehensive, vertiginous kind of sexual experience, but he exemplifies and captures some of the crucial points of my discussion so far when he refers to an "angoscia" whose origin may actually derive from the time-bound perishability of body and desire.[94]

It is, however, the woman's wished-for solution to her situation of sexual suffering and anxiety (first just imagined and then, at the conclusion of the story, nearly actualized) that adds further consistency to the alleged presence of certain aspects of the "sex-appeal of the inorganic" in Marinetti's erotic short fiction:

> Caro! Caro amore, perché non vuoi che io ti beva il sangue? Sogno di aprirti una ferita piccina piccina e con un piccolo tubetto d'oro suggere il tuo sangue a piccoli sorsi. Oh! Non temere. Non ne morrai [...] Dammi il tuo sangue da bere. Ti porterò in me [...] Vorrei strapparti la pelle. [...] Ti voglio così nudo, scorticatissimo! Svestiti della tua pelle! (218–19)

The "biondina's" urge to drink all of her lover's blood and peel off his skin so that she may reach his inner recesses, in fact, begs to be interpreted in the light of some of Perniola's observations on the sexual nature of vampirism:

> Il vampiro beve, prosciuga, secca il corpo di chi ha commercio sessuale con lui: è come se non solo il seme, ma tutti i liquidi fossero assorbiti nella sua bocca [...]. Nella fantasia che tutto il nostro sangue sia assorbito dal suo bacio, prepotentemente urge e imperiosamente s'impone il grande richiamo dell'inorganico. Se il corpo della nostra amante fosse davvero liberato dal sangue, il suo interno diventerebbe una grande tasca [...] esso sarebbe finalmente idoneo ad una penetrazione illimitata (Perniola 1994, 101–2).

Although Perniola's point of view is that of a male lover, the situation he describes seamlessly adapts to the fictional core and to the female protagonist in "La guancia."[95] What is, in fact, the true objective of her fantasy

94 In this sense, a title such as Cangiullo's *L'amante che non morrà* (1919) is also quite indicative.

95 On the other hand – substituting vampirism with fetishism and necrophilia – Perniola's words may be quite appropriate and adapted also to describe "La carne congelata," a story which represents, this time from a male perspective, a similar "inorganic" situation of objectification and transformation (of a woman) into a

if not stripping her lover naked until he loses his "organic cloth"? What is her goal other than making him a mere aggregate of genderless, palpitating blood and flesh – a neutral, "live thing" ("non ne morrai"), "senza soggetto, senz'anima e senza corpo vitale" (Perniola 1994, 61), which will never be able to leave her using his "gambe giovani" and which, instead, will be always available for her unlimited possession? Additionally, what does her lover's body become in her imagination if not a "tessut[o] della [...] carne" which can be "worn" or "put on" ("ti porterò in me"), where "muscoli" are "cordami," "nervi" are "capigliature" and "fibre" have a taste (*Novelle* 219), and whose major function, exactly like the inorganic "tessuti carnali" mentioned by Perniola, is that of "coprire e [...] essere coperti, [...] vestire e [...] essere vestiti, [...] avvolgere e [...] essere avvolti" (Perniola 1994, 61)?[96]

In this scenario, it is only natural that the bloody "tasca di otto centimetri bene aperta [...] con risvolto" that, at the conclusion of the story, opens up on the "guancia destra" of the man because of his street fight and that the "biondina" is eager to tend, is a "magnifica ferita" which, most appropriately, "sembra tagliata da un sarto sapiente" (221). As such, and as the additional clothing lexicon indicates, it suggests once more both her desperate urge for an "esperienza della pelle e del corpo come insieme di tessuti" (Perniola 1994, 61) and, as a direct consequence, the location of her deepest sexual gratification in a quintessentially inorganic experience.[97]

---

thing-like "pezzo di carne [...] che non fugga mai" (262–3), used to satisfy the male protagonist's sexual desire.

96 Marinetti's declared intention in "Il tattilismo" to "penetrare meglio [...] la vera essenza della materia" (*TIF* 179) and his attempt to eroticize the whole body to the detriment of genital sexuality come also to mind in these circumstances.

97 According to Perniola, philosophy has the task of removing "sexuality from its vitalistic [...] excesses" in order to reach the objective of "a virtual, neutral sexuality which is neither precarious nor ephemeral as organic [...] but always available in its vertiginous artificiality" in *Ritual Thinking. Sexuality, Death, World* (New York: Humanity Books, 2001), 35–6.

It may be worth noting, in conclusion, that this "path towards the inorganic" that I attempted to trace here in the limited context of these erotic stories may anticipate experiences and themes that are more frequently associated with a postmodern sensibility and (much more unexpectedly) even contemporary eco-critical theories. I am thinking, for example, at Jane Bennett's *Vibrant Matter. A Political Ecology of Things*, in which she contends that "there is also public value in following the scent of a nonhuman, thingly power, the material agency of natural bodies and technological artifacts". Her book, in other words, seems to "futuristically" aim at blurring the lines between inert matter and vital life in order to "highlight the active role of nonhuman materials (including "edible matter") in public life" and, at the same time, well exemplifies the "material turn" of recent philosophical thought.[98]

In the next section of this chapter, we shall see that such a "path" – (an ideal solution to futurism's impasses in a variety of matters: time, desire, sexuality, and morality) – continues to wind up, often signalled and accompanied by alimentary references, also in those additional texts by Marinetti and other futurists in the Thirties in which the familiar, "old" quest to achieve a "mechanized man" is informed by new technologies and sensibilities.[99]

98 Jane Bennett, *Vibrant Matter. A Political Ecology of Things* (Durham and London: Duke University Press, 2010) xiii, 2. The author states that her methodology reflects a "materialism in the tradition of Democritus-Epicurus-Spinoza-Diderot-Deleuze more than Hegel-Marx-Adorno" (xiii).

99 As an example of the "hard-to-die" presence of alimentary references (also) in connection with the drive towards the (inorganic) "mechanization of man," see the following passage in *Simultanina. Divertimento futurista in 16 sintesi* (1930), in which the character of "Il Buongustaio" addresses the audience at the beginning of the second "sintesi" by saying: "A Simultanina non ho mostrato il mio panciotto simultaneo ... brevettato ... 11 scompartimenti. Contiene tutto: mangiare bere digerire scrivere sparare. Serve anche da corazza contro l'amore." (In Giovanni Calendoli, ed., *Teatro. F.T. Marinetti*, Vol. III. Roma: Vito Bianco Editore, 1960), 393.

1.4 "Sweet Steel": the "morale antropofaga" in Marinetti's *Patriotismo insetticida. Romanzo d'avventure legislative*

Given the proliferation, especially in the last couple of decades, of studies which address the relationship between nationalism and modernity in the Italian avant-garde[100] and the widely shared opinion among scholars of futurism that "nationalism or patriotism (complete dedication to the Fatherland) is to be taken as the fundamental motivation of Marinetti's cultural and social activity,"[101] it seems a little surprising that a text such as Marinetti's *Patriotismo insetticida. Romanzo d'avventure legislative* (1939) (from now on *Patriotismo*) has been barely addressed in a critical context.[102] One reason for this lack of attention – besides the understandable unwillingness of scholars to deal with obscure works of far-from-stellar literary quality and often perplexing, objectionable content – may be partly due to issues of perceived importance and periodization (i.e., the tendency to

100 See, most notably, Emilio Gentile, "The Conquest of Modernity: From Modernist Nationalism to Fascism," *Modernism/Modernity* 1.3 (1994), 55–87; Günther Berghaus, *Futurism and Politics. Between Anarchist Rebellion and Fascist Reaction, 1909–1944* (Providence: RI: Berghahn Books, 1996); Marja Härmänmaa, *Un patriota che sfidò la decadenza: F.T. Marinetti e l'idea dell'uomo nuovo fascista, 1929–1944* (Helsinki: Accademia scientiarum fennica, 2000); Walter Adamson: *Avant-garde Florence: From Modernism to Fascism* (Cambridge, MA: Harvard University Press, 1993); Andrew Hewitt, *Fascist Modernism: Aesthetics, Politics, and the Avant-Garde* (Stanford: Stanford University Press, 1993).

101 Marja Härmänmaa, "Beyond Anarchism Marinetti's Futurist (anti-)Utopia of Individualism and 'Artocracy'," *The European Legacy*, Vol. 14, n. 7 (2009), 857–71, (865).

102 F.T. Marinetti, *Patriotismo insetticida. Romanzo d'avventure legislative*, in Giovanni Calendoli, ed., *Teatro*, Vol. I (1960). This "romanzo," as Calendoli notes, is a modification of an earlier text entitled *Il Club dei simpatici* (Calendoli 281). Calendoli, by the way, in his "Introduzione," mentions and briefly discusses every theatrical work by Marinetti included in his edition. The only one missing from his analysis is *Patriotismo*.

segment literary studies into periods). Emilio Gentile's influential article on the relations between futurism and fascism, for example, by linking his notion of "modernist nationalism" to "a state of mind, a sensibility, a cultural orientation centered on the myth of the nation, a sensibility that we can find in all the avant-garde movements formed in Italy during the fifteen-year period prior to the Great War" (59–60), focuses on the earlier and more radically innovative part of futurist history (1909–1920). Thus, although he obviously underlines the continuities between such "sensibility" and the "fascist modernism," his focus is historical and does not consider specific literary manifestations of the nationalistic/patriotic discursive formation in the later years of "aero-futurism" to which *Patriotismo* belongs. Similarly, Hewitt in his volume *Fascist Modernism* ignores the "later developments (or degeneration) of 'secondo futurismo' ('aerofuturism') perhaps because this phase may be seen as an outworking of Marinetti's primary phase."[103]

As far as I can tell, Marja Härmänmaa is the only scholar who in her book *Un patriota che sfidò la decadenza*, devotes some comments to *Patriotismo*. She associates this allegorical text with a group of writings from the Thirties (i.e., *Luci veloci*, *Simultanina*) in which Marinetti repeatedly addresses the Nietzschian question of the "relatività della morale, dei concetti del bene e del male, e della loro dipendenza dal tempo e dalle circostanze" (70). After noting that, for Marinetti, "L'imperativo di assoggettare tutto alla patria aveva molteplici conseguenze e influenza anche la concezione della morale" (69), Härmänmaa provides a summary of its first chapters, delineating the book's main plot ("il protagonista Paranza, 'una sferica rosea caldaia,' deve processare alcuni membri della società misteriosa dei Lungimiranti che hanno commesso vari delitti,") (72) and concludes – echoing Marinetti's own words in the preface of the book – by reasserting that the "conseguenza di questa nuova morale era che tutto ciò che veniva fatto per il bene della patria era lecito e accettabile" (73). *Patriotismo* is then mentioned again in a different section of her study, when she discusses

103 Stanley Payne in his review of Hewitt's book, *Modernism/Modernity* 1.1 (1994), 160–1.

Marinetti's condemnation of "esterofilia" which – she observes – "insieme alla questione della morale, è uno dei temi principali nel dramma" (258).

In both the preface and the main text, Marinetti (despite his original 1912 ban on adjectives in the "Manifesto tecnico") describes such a "morale" with an abundance of familiar, futurist qualifiers: thus, it is "nuova," "simultanea," "ottimista e futurista," "aerea," "vasta," "calda," "profumata," and opposed to the "old" one born out of "Mondialismo umanitario egoismo individuale meschino mascherato di solidarietà astratta e teorica" which, instead, was "pessimista nostalgica e vile di anarcoidismo e parole vuote" (92). However, notably, the last adjective that defines how this new "morale" should be, just before the conclusion of the text, is of a completely different sort and, given the focus of this study, is the one that caught my attention. It appears in the penultimate (thirteenth) short chapter entitled "La filosofia dei denti," in a section where, if possible, the presence of alimentary images is even more pervasive. Here we find the character of Curreno, one of the cannibals belonging to the local Tucrù tribe, whom Paranza and his improbable group of travel companions encounter when they arrive after a twelve-hour flight by hydroplane to the "Isola gastronomica a prezzo fisso" in the Fiji archipelago. Curreno is swimming in the "acqua turchina con vene verdi e violacee" (193) of a bay in the island, showing his "crocodile teeth," towards the anchored plane which is shaking and jolting "inesplicabilmente nell'acqua immobile." When he reaches it, he exclaims: "Stiamo allargando alle machine europee la nostra morale *antropofaga* (italics mine) Addenterò dopo le ali il carburatore che spero trasmetterà la sua capacità vorace nel mio stomaco imperfetto" (194). Following Curreno's puzzling statement, Tokkamatok, the Tucrù's chief, then asks Paranza – who, in the meantime arrives and witnesses the scene – his opinion about his own intention to "mangiare lo stabilizzatore per equilibrare il mio corpo che oscilla troppo quando bevo sangue fermentato" (194). As we discover, in fact, Tokkamatok had already arrived at the plane and was already inside it, "lavorando coi denti nella carlinga" (194). To complete the scene, at the beginning of this same episode, we learn from the stage directions provided by the narrator that Paranza himself "Da parecchi giorni [...] aveva adottato la nudità fogliuta e pennuta che costituiva l'abito cannibalesco di parata" (193). Upon his arrival at the island, when he and his fellow travelers find out that the cannibals would just like to receive "Liiiiibri da mangiare cogli occhi" (180) as a gift ("un piccolo ricordo d'Europa") (179), Paranza's reply is that: "Disprezzavo tanto i libri alla partenza che li ho tutti rimpiazzati con belle dentiere d'avorio prevedendo una nostra trasformazione in cannibali" (180).

Now, leaving aside any attempt to interpret these strange events for the moment, the adjectival choice referred to the "morale" ("antropofaga") may not be particularly surprising at all. First, it is literally coming out of the mouth of a cannibal and at the conclusion of a quintessential pastiche,[104] both from the point of view of form and content, which is constituted by fourteen, loosely connected scenes mostly stuffed with gastronomical references. Second, as we should by now know, the adjective is the product of a writer certainly familiar with the allegorical potential of alimentary images from his earlier pre-futurist phase, from *Le Roi Bombance* on.[105] Indeed, especially after the "lesson" of *Le Roi Bombance*, the appearance of the trope of cannibalism and, in general, of a variety of gastronomical references in this later context may invite further reflection about their allegorical relevance and suggest an interpretation of *Patriotismo* within the cultural and political coordinates of the late 1930s, a period marked by

104 In Marinetti's intentions *Patriotismo* was supposed to be a "romanzo sintetico simultaneo lirico teatrale e cinematografico da mettere sul palcoscenico o sullo schermo" (92).

105 In both *Le Roi Bombance* and *Patriotismo* the alimentary discourse needs to be read in allegorical terms and as a means to address and comment on social, political, and aesthetical questions. Thus, despite all the differences between the two texts, I cannot help but thinking of the existence of an ongoing dialogue between these two texts, written thirty years apart. After all, as James Joyce once said, "all novelists have only one story, which they tell again and again" (qtd. by Joan Acocella, *The New Yorker*, 16 May 2011, 120). On a completely different level, and mostly just as a curiosity, *Patriotismo*'s last section shares the exotic setting, the cannibalistic theme, and the (wishful) cheerful dramaturgy with Sergio Tofano's 1936 musical comedy, *"L'isola dei Pappagalli. Bonaventura prigioniero degli antropofagi."* Although aimed at children, Tofano's text is not alien from the influence of futurist poetics and its story too – with Bonaventura's and his friends' failed conquest of the island – ironically and critically refers to Italian colonial adventures in the Thirties.

increasing tensions between futurism and an increasingly invasive fascist cultural bureaucracy.

If, in other words, in *Le Roi Bombance* "il tema dominante [...] è [...] quello [...] della fame, sovrana del mondo [...] principio e fine di ogni realtà [...] [e il testo] denuncia lo scetticismo con il quale F.T. Marinetti accolse anche l'ideologia del Socialismo [...] distrugge tutti gli ideali ed anzitutto quello della politica'" (Calendoli 1960, ix/x/xvi), one may wonder if, more than thirty years later, a comparable attempt to decrypt the recurrent alimentary discourse in *Patriotismo* – which, if we agree with Paranza when he says, "Ma occorre capite occorre mangiare" (188), is equally pervaded by the theme of hunger – could provide further insight not only on its explicit theme (i.e., the need to expand one's moral judgment in the name of patriotic interests), but also additional information on the confusion and discrepancies between Marinetti's aesthetic and ideological position towards the end of his literary career and the contemporary fascist one.

As Berghaus summarizes, "the years 1923 to 1930 were characterized by [Marinetti's] often desperate desire to gain recognition by the new regime" to the extent that "in 1924, he issued a manifesto, *Le futurisme mondiale*, that portrayed futurism as a movement of significant influence on the international avant-garde" and, later on, the collection of essays, *Futurismo e Fascismo*, which "presented Futurism as an integral element of Fascism's rise to power." The 1930s, however, were a time when futurism basically struggled to survive in "the increasingly stifling political climate of the peninsula [...] Marinetti fought against the concept of degenerate art, against the burning of books and works of art, against political censorship and anti-Semitism," until "in January 1939 [the fascist hard-liners] shut down the Futurist paper *Artecrazia* and, with it, Futurism as a viable artistic movement" (366). On this historical background, I would like to point out two things: first, that the preface of *Patriotismo*, signed by "F.T. Marinetti Sansepolcrista" is dated just eight months after this "shutting down," "August 1939–XVII." Second, the ideological basis of the origin of

"Sansepolcrismo"[106] could be summarized in Mussolini's statement: "Noi ci permettiamo di essere aristocratici e democratici, conservatori e progressisti, reazionari e rivoluzionari, legalisti e illegalisti, a seconda delle circostanze di tempo, di luogo e di ambiente."[107] Now, *mutatis mutandis*, it is tempting to hear an echo of the arguably "relativistic ideology" underlying the Duce's words in Paranza's controversial remarks in which the question of "cultural/moral relativism" emerges more clearly. Thus, for example, when he states: "Bisogna trasformare e allargare il codice Per esempio in certi casi l'antropofagia è difendibile Ah il mio sogno di visitare i cannibali delle isole Figi" (125), or, similarly: "La velocità ha la sua morale Il patriottismo ha la sua La Guerra ha la sua Tutte diverse spesso opposte Questo idrovolante m'impone la sua d'acciaio dolce" (164), he not only sounds like a "chief legislator" himself, but he also appears to be artistically and "futuristically" interpreting and rephrasing an important principle of Mussolini's original, fascist political program around the time of Sansepolcro. And given the history of Marinetti's relation with the regime and its leader[108] and the frequent public support that futurism gave to fascist policies (even though often fundamentally disagreeing with them) this sort of artistic (and implicitly propagandistic) reinforcement of key fascist principles is certainly not anomalous.

106 "Sansepolcrismo" refers to the original Fascist programme which was presented in Milan's San Sepolcro square on 23 March 1919. It was then published within the *Manifesto dei Fasci italiani di combattimento*, in *Il Popolo d'Italia*, 6 June 1919. Interestingly, the second point of the programme, ambiguously reads: "II. The meeting of 23 March declares itself opposed to the imperialism of other peoples to Italy's detriment and to any possible Italian imperialism to the detriment of other peoples. It accepts the supreme postulate of the League of the Nations which presupposes the integration of each nation, an integration which in Italy's case must me realized on the Alps and on the Adriatic with the claim to Fiume and Dalmatia and their annexation." In Paul O'Brien, *Mussolini in the First World War: the Journalist, the Soldier, the Fascist* (Oxford: Berg, 2004), 13.

107 Benito Mussolini, *Dopo due anni*, in *Il Popolo d'Italia*, 23 marzo 1921, then in *Scritti e discorsi*, vol. II, Milano (1934), 153.

108 On this relation, see Härmänmaa, *Un patriota*, 291–303.

With regards to the specific reference to anthropophagy in Paranza's remark, Marinetti's resort to the trope of cannibalism and the frequency of gastronomical images and metaphors is probably due also to the artistic necessity of trying to spice up a subject-matter (i.e., a "romanzo di avventure legislative") that otherwise could be dangerously boring for his audience.[109] As a consequence, the attention-grabbing reference to the legitimacy of the sin of anthropophagy contributes to emphasize the main message regarding the need to "variare continuamente i limiti del male e del bene," to expand the moral limits of the code of law, and advance the idea that "tutti i peccati po[tevano] meritare qualche indulgenza [that is, even anthropophagy] eccettuato uno che [era l'] antipatriottismo esterofilo [...]" (in Härmänmaa 258), while less outrageous, and yet equally memorable, culinary references (Paranza: "Hai pensato Condirina che dalla tua nuova cucina deve uscire una calda profumata morale nuova la quale allarghi benignamente il detto popolare il fine giustifica i mezzi" [192]) keep reinforcing such a message throughout. Although the characters themselves (Paranza above all) often speak through elaborate images of eating and food,[110] the stage directions are the sites where Marinetti, as narrator/director, can truly reveal his eminently alimentary perception of reality. The beginning of chapter five, for example, reads:

> Un'aurora gastronomica/Cielo pomidorato/La villa del giudice Paranza è fortunatamente un cubo di dolce ricotta al fresco nelle vigne bene disposte nella campagna/Un grande lago di vero latte appena munto/L'ha però scremato l'ultimo fiato benigno della note che volle ungerne le nostre labbra bruciate dai sogni/Sapore squisitamente zuccherino e vanigliato dell'aria/Essere saporito/Per chi?/Forse per fantasie cucinarie di quella brezza affaccendata tra nuvolose caldaie di bollito d'oro/Dentini

109 At the same time, reflecting on this subtitle after I happened to watch (the millionth?) episode of the American TV show "Law and Order," I realize that "boring" may be an inaccurate term. Marinetti could have already intuited the immense media potential and appeal of the legal drama!

110 See, for example, what Paranza says when Urò, a fellow magistrate, initially notes that they are in the "Archivio dei Grandi Processi Misteriosi": Paranza: "Ciò mi spiega la ribalderia dei topi e la loro ghiottoneria fra tante leggi inacidite delinquenze svaporate e cartacce giuridicamente saporite" (93–4).

> diacci di un ramo che mi sfiora/Intorno alti torroni mandorlati delle montagne/Da mordere più tardi dopo la marmellata di visciole del mare/Ora mentre l'atmosfera spreme miliardi di chichi d'uva moscato color oro vecchio giunge in bocca tutto un bosco di fresca insalata indivia lattuga con olio di sole e aceto di brina/La cucina di villa Paranza splende di stelle comete caschi e aureole cucinarie sospese ai muri abbacinanti [...] (121).

A comparable culinary description of the landscape appears then in chapter 10, the one dealing with the previously mentioned, cannibal-infested "isola gastronomica a prezzo fisso." Here however, the emphasis is expectedly on the "meat" rather than on the "salad":

> Gioia aviatoria di Paranza/Chi ha fame di cielo sostanzioso?/Oggi tutto carne/Alta macelleria del tramonto equatoriale/Le nuvole squartate sono sospese per le zampe posteriori a ganci di diamante/Quelle prime stelle nel verde putrescente dello zenith/Giù i musi fumano e vomitano sangue sull'isola polputa chiazzata di viola e porpora/Ossame e budellame della foresta di liane camerus palme baobab mango e fluttuanti materassi di piante parassite [...] (171).

Regardless of the alternatively vegetarian or carnivorous emphasis, what seems especially worth noting in these passages, from a rhetorical and stylistic perspective, is the simultaneous combination of the high and the low, that is, of the horizontal, earthly language of food and the vertical, heavenly one inspired by the experience and the perspective of flight. Thus, expressions such as "cielo pomidorato," "cielo sostanzioso," and "nuvole squartate" convey not only a familiar reading of landscape in culinary terms and the attempt to stir several senses at once, but they also have the (involuntary?) effect of never letting these poetic images go too far away from the ground and from a more tangible (or edible) reality; that is, preventing them from veering towards the "literary insipidness" that could derive from "flying too high" (i.e., merely using a rarefied, cosmic language). Of course, one may find this situation a bit paradoxical, since this tendency to keep at least a rhetorical tether to the earth cohabits with the opposite wish to leave it behind (or below) to reach the heights of a new, true spiritual and moral elevation. As Härmänmaa noted, "per Marinetti la passione per il volo nasce dalla volontà di sfuggire dalla taccagneria e dalle pedanti minuzie di tutto ciò che è terrestre" (141). The fact that, in the end, the hydroplane itself becomes food for the cannibals seems to capture well this contradictory dynamics by transforming that which is supposed to fly high into "low matter" so that, in turn, hypothetically, even the foreign, exotic primitives will be able to access the same (moral, patriotic) heights of the (western) "Lungimiranti." But I shall examine this scene in further detail later on.

For the moment, keeping in mind the importance of this sort of alimentary rhetoric in defining patterns of thought, I also wish to point out that in a text whose main declared objective is to redefine the commonly accepted idea of "morale" through the "Adorazione concreta della Patria" and which, as I have noted, is apparently sharing and voicing an original (sansepolcrista) fascist perspective in such redefinition, a link seems to be established, at least at the imaginary level, between the "anthropophagous moral" on the one side, and, on the other, the much more concrete expanding borders and assimilating body of the Italian fatherland at the time. Notably, the publication of *Patriotismo* follows the fascist colonial enterprise in Africa (Ethiopian War 1935–6), Mussolini's declaration of the "Italian Empire" (1936) and his widespread dreams of a "Greater Italy," and the successful annexation of Ethiopia and the Italo-German Pact of Steel (22 May 1939). Significantly, in the preface, the terms "espansione" and "perfezionamento" are in the same paragraph with "Individuo" and "Patria." As Marinetti puts it, "la fusione dell'Individuo con la Patria divina costituisce l'unico egoismo nobile perché vasto concreto razionale e lirico nella sua continua espansione e nel suo continuo perfezionamento" (92). Marinetti had already suggested the existence of an almost physiological connection between the individual (to whom the moral sphere necessarily pertains) and the (divine) fatherland when, in "Patriottismo futurista" (in *Democrazia futurista*, 1919), he had unequivocally stated that "la patria è il massimo prolungamento dell'individuo o meglio: il più vasto individuo vivo capace di vivere lungamente, di dirigere, dominare e difendere tutte le parti del suo corpo" (*TIF* 390).[111] In this light, one more theme implicitly

111 With regards to the relationship between body and nation in this period, Victoria De Grazia aptly reminds us that "the Duce equated the health of individuals with the well-being of the nation" in V. De Grazia, "Nationalizing Women: The Competition

reflected in *Patriotismo*, especially in its conclusive exotic section, is the kind of relationship that the extended main body of the fatherland has with its other parts. Or, to put it differently, the inherently colonial/cannibalistic question of how one should literally and metaphorically feed this "corpo" (i.e., assimilating human beings as food and, in turn, the land where these human beings live) in order to make it "live a long life [...] manage [and] dominate" all its appendixes, does not seem completely irrelevant or alien in *Patriotismo*, in particular if read in the background of the fascist ongoing redefinition and expansion of the term "patria" into "empire."

This said, after having individuated some instances in which *Patriotismo*'s main ideological message echoes an "original" fascist one, in terms of the absolute love for the fatherland and the need to vary one's moral perspective according to time and place, I cannot but wonder if this picture actually captures the whole flavor of the text. For example, what may it tell us about the contradictions and the differences between Marinetti's aesthetic/ideological position and fascism's ideological position in the 1930s a text in which the declared objective of nurturing the "concrete" love of the Italian fatherland – a goal shared by both futurism and fascism – is eventually dramatized outside of Italian territories, among cannibals on a Fiji island reached after a flight on a hydroplane piloted by a homosexual named "Antidonna"?[112] How significant is it, in the light of the

between Fascist and Commercial Cultural Models in Mussolini's Italy," in *The Sex of Things; Gender and Consumption in Historical Perspective* (Berkeley and Los Angeles: University of California Press, 1996), 337–58, (344).

112 Although Paranza's attitude towards Antidonna and his sexual orientation during the flight is not what today would be called politically correct ("Voi [...] siete la menzogna l'odio-amore dello scandalo il desiderio-piacere del pericolo La vostra estetica di pizzi vaporosi e moine vi fa odiare l'utero crudo ghiotto della donna Siete omosessuali in quanto non vi dichiarate mai tali [...] Che strano equipaggio mi è capitato Un pilota quasidonna e un motorista piroettante [Piroetta]" (166), his is a position of clear tolerance that is certainly far from the criminalization homosexuality will eventually face under fascism (with repressive measures such as "domicilio coatto" and confinement). As he tells Antidonna, "Ad ogni modo non voglio disturbare il tuo esperimento personale con delle vane condanne caro Antidonna" (162). Eszter notes that "The Rocco Penal Code [...] did not contain a specific anti-homosexual

battle Marinetti conducted in these years against the burning of books and works of art, against political censorship, that these cannibals are hungry for books and that their "mascelle meccanizzate" sound and even look like "macchine per scrivere"?[113] Furthermore, how relevant is it, in the light of Marinetti's combined condemnation of "esterofilia" and interest in "legislative adventures," that the one law book given to the cannibals as a gift is not the recently promulgated new fascist Penal Code, the 1931 Rocco Code (so-called after Alfredo Rocco, the fascist Minister of Justice who sponsored it) which certainly contributed to create a distinct fascist legal (penal) culture but, instead, "il libro piú importante d'Europa un piccolo codice napoleonico tascabile Tradotto in inglese" whose interpreter is a cannibal who states "Sono genovese ma conosco l'inglese Ho lavorato in Rue de la Bourse e in Wall Street Leggerò e commenterò io il codice al popolo cannibalesco" (181)?[114] More importantly, how should one interpret the fact that, in the end, not only Paranza "goes native" (reminiscent of Conrad's Mr. Kurtz in *Heart of Darkness*) by adopting the costumes of the Fijian cannibals, but also that Curreno and Tokkamatok, both savages belonging

law, and neither did the previous Zanardelli Code. However in both cases, the decision to omit such a law from the Penal Code was not the result of liberal considerations. The reasons for not introducing the anti-homosexual article into the Penal Code can provide useful indications how Fascism treated homosexuality. Cover up strategies, occultation, and negation of the fact that homosexuality existed in Italy were significant factors to eliminate any kind of discussions on it. The political power utilized this strategy of silence which Dall'Orto calls 'repressive tolerance.' The predilection of the State was not the fight openly against homosexuality but to operate, and repress in silence." Eszter Andits, "'Sore on the Nation's Body': Repression of Homosexuals under Italian Fascism," unpublished thesis (2010), 66 www.etd.ceu.hu/2010/andits_eszter.pdf. [Last accessed July 2012].

113 "In realtà il continuo e veloce funzionamento delle due dentiere o machine per scrivere sovrapposte aveva forzato e annullato labbra occhi fronte e capelli" (182).

114 The entry "Codice Rocco" in the *Encyclopedia of Contemporary Italian Culture* (ed. by Gino Moliterno) notes that "its architect was a leading authoritarian nationalist intellectual who became a prominent state technocrat, and the Rocco code was undeniably repressive of associational and individual rights, not least those of women. The code aimed above all to defend the state, and while major amendments were made in 1944–5, it was also used to repress opposition in the new democracy."

to a primitive, foreign civilization are lastly seen as chewing away at the hydroplane, one of the most powerful symbols of the futurist imaginary and its current project of "aeropoesia"? Why is that one of those machines which both inspired and contributed to boost the futurist project and poetics in this later "aerial" phase, "an image that would symbolize Mussolini's government" (Bohn 216), one crucial to understanding futurism's position in fascist Italy, and that certainly helped, with its essential role in various military enterprises, to increase Italians' sense of patriotism, ends up being devoured by two cannibals with the silent "blessing" of a plump magistrate who is more interested in "processing" the foods prepared by Condirina, his female cook (97), than in responding in any way to Tokkamatok's final query and gesture?[115] Finally, would it be possible to tentatively suggest that *Patriotismo* is also ambiguously and contradictorily commenting on the tensions that a group of truly patriotic individuals belonging to a weirdly familiar, futurist-sounding "Società dei Lungimiranti" is having with an increasingly voracious culture and a trial-prone, authoritarian, and repressive "società" led by a "capo" who, imaginatively overlapping with the caricature of Tokkamatok, may well be cheerfully described as having a "testa sproporzionata" and a "bocca simile ad una botola"?

The best way to begin addressing some of these questions is to return to the text and further reflect on some of the potential implications of the emblematic scene depicting Curreno's and Tokkamatok's "consuming" of the plane.

115 "Masnata published a poem in October 1934 entitled 'Fascismo: aeropoesia in parole in libertà.' Seeking an image that would symbolize Mussolini's government, he chose a powerful hydroplane. Ever since Italo Balbo crossed the Atlantic with a group of flying boats (in 1930 and again in 1933), the seaplane was viewed as the airplane of the future. Not only was it capable of long-distance flights, but it could land or take off almost anywhere. [...] Since Italy essentially found itself on a war footing, aeropoetry became more and more preoccupied with aerial warfare. Aeropoems increasingly dealt with war planes, bombardments, and aerial battles." Willard Bohn, "The Poetics of Flight: Futurist 'Aeropoesia,'" *MLN*, Vol. 121, Number 1 (2006) (Italian Issue), 207–24.

First of all, there are few doubts that this gesture needs (also) to be read in the light of the familiar and yet updated desire also among several futurists in the Thirties to become identified with a machine, to achieve the inorganic and time-defying efficiency of a well-oiled engine, be it that of an automobile's or an airplane. Although I agree with Härmänmaa when she observes that "l'uomo nuovo di Marinetti, negli anni Trenta e Quaranta, non era più un Superuomo meccanizzato. In *L'Aeropoema di Gesù* (1943–1944) egli rivela che l'identificazione dell'uomo con il motore è solo un sogno mai realizzato né realizzabile," (285) it seems also true that in *Patriotismo* such a utopian fusion is still present, fantasized, and desired. Besides the episode under consideration, one may look also at the description of the "chemical," erotic attraction between the various dancing partners in chapter 8 ("Il poema elettrochimico di una festa da ballo in casa Paranza") to find a variation of this persistent longing for the ideal of "metallized flesh": ("Lady Delta è un asse d'acciaio a vite con volantini [...] L'affinità dei due metalli umani solubili allo stato solido vuole formare un composto intermetallico o figlio Ma dove sono le pinze termoelettriche delle labbra? [...] Già la saldatura dei due metalli-carne aspetta ordini [...]" [154–5]).

At the same time, as I briefly suggested before, the "perfecting" and "balancing" physiological effect that Curreno and Tokkamatok are hoping to achieve by ingesting specific parts of the hydroplane also alludes to the implicit superiority, the healing and, thus, civilizing powers of western, colonial technology and culture. Beyond its outrageous absurdity, the idea behind Tokkamatok's explanation to Paranza about the logic of his eating missionaries (2) "Il miglior modo di convertirci è quello di ingoiare il convertitore amalgamandone così la fede," (187)[116] is the same one that also informs his consuming of the plane's "stabilizzatore" or Curreno's ingestion of the "carburatore." By eating these inanimate objects, the cannibals perform a sort of degraded rite of communion, one which allows them to fetishistically assimilate the objects' inherent futuristic/fascistic, western qualities and, thus, to allegedly improve their existential and spiritual status.

116 In Italian "convertitore" may refer not only to "the person who converts" but also to a mechanical (electric) part in an engine ("convertor"). Of course, the pun is intended.

The radical and perhaps unexpected novelty in this techno-primitive scenario is, however, that the reverse of this dynamic is said to have positive effects. That is, the expansion of the cannibals' "morale antropofaga" to the "macchine europee" is not seen, as one might expect given the preeminence given to patriotic, nationalistic sentiments, as something negative, a sort of "infectious disease" coming from a foreign Other that could weaken the love for Italy but, rather, as a "key ingredient" that is going to enhance precisely one's patriotism as it simultaneously redefines the notions of good and evil. As a matter of fact, in this instance, Marinetti seems almost to play with, echo, and adapt to the Italian situation Oswald de Andrade's anthropophagic project in Brazilian *modernismo*.[117] As Sérgio Luiz Prado Bellei observes, "Andrade saw Brazilian society as paradoxically divided between two worlds, one primitive and backward, the other bourgeois and modern. But while to that date in Brazilian history primitivism and backwardness had always been a source of embarrassment, Andrade's *antropofagia* proposes [...] to treat primitivism and backwardness optimistically as a positive value."[118] In his *Manifesto Antropófago* (1928), he argued that "cannibalizing" other cultures was a way for Brazil to assert itself against European postcolonial cultural domination. Although, from a colonial perspective, there are obvious differences between the status of Italy and Brazil, as there are between Marinetti's and de Andrade's ultimate objectives and attitudes towards primitivism, it is also true that the founder of futurism had a comparable vision of his own Italian society as split in two, between backward "passatisti" and modern "futuristi." He was similarly embarassed by the civic/patriotic apathy that affected Italians and the "senso di inferiorità di cui gli italiani [...] soffrivano nei confronti delle altre nazioni" (Härmänmaa 257). If for Andrade primitivism, emblematized

117 Marinetti's Brazilian trip took place in 1926; however, he did not meet Andrade on that occasion because he was abroad. See Jeffrey Schnapp; João Cezar de Castro Rocha, "Brazilian Velocities: On Marinetti's 1926 Trip to South America", *South Central Review*, vol. 13, n. 2/3 (Summer–Autumn, 1996), 105–56.

118 Sérgio Luiz Prado Bellei, "Brazilian Anthropophagy Revisited," in Francis Barker et al., eds, *Cannibalism and the Colonial World* (Cambridge UK: Cambridge University Press, 1998), 87–109, (103).

by cannibalism, was eventually to be "understood [...] as a valuable instrument for redeeming modern society from the excesses of capitalism" and, therefore, coincide with "emancipation" (Prado Bellei 102), for Marinetti the primitive "morale antropofaga" which devours the "macchine europee" and, implicitly, the foreign cultures and discourses which define them, becomes "positively charged" – the necessary instrument to stir Italians, unite them around a common patriotic cause, and boost their sense of national cultural identity.

At this point, let me say that I realize my last set of observations on this episode is difficult to support with strong textual evidence. However, given the Italian cultural and political climate at the time of the publication of *Patriotismo* and the accumulation of my remarks so far, I cannot help but suggest the existence of some kind of relationship between the fictional phagocitation of the hydroplane, with all its symbolic baggage as an emblem of futurist (aero)poetics in these years, and the concrete blow that fascist hard-liners gave to futurism by shutting down *Artecrazia* only a few months before *Patriotismo* appeared in print. The fact that, as noted earlier, the hydroplane may also be simultaneously representative of fascism's military prowess should not necessarily undermine this suggestion. The implication is that if things remain the same and the current penal/legislative code is not revised, thus implicitly recognizing the importance of futurism's contribution to the "sorti della nazione," fascism itself could end up negatively affected and dismantled piece by piece in a similar way.

I find it intriguing and typical of futurism's oscillation between "una retorica del materialismo e una retorica della spiritualità" (Valesio 251), that while this scene contains a constructive side that is supposed to expand and enhance an old concept of morals, it is also clearly destructive in its essence and, as such, it unavoidably conveys also a sensation of fatality, of an ending. The hydroplane seems at once a source of spiritual inspiration, a superfood able to transform and improve the status quo along futurist directives, and also a big, trembling metal insect, perhaps put down by the same "insetticida" of the title, weakened and powerless in the face of its imminent, sacrificial disembodiment. Additionally, given the circumstances, it is also tempting to view the hydroplane as a potential example of that association of metal with the notion of a vital materiality that Deleuze and Guattari discuss in a brief section of their "Nomadology" plateau. In particular, their intuition that "metal best reveals [a] quivering effeverscence; [...] bursting with a life that gives rise to 'the prodigious idea of Nonorganic Life'" (qtd. in Bennett 55), seems to resonate well with the situation depicted in this context, in which steel is consumed as if it were an organic matter affecting both private and public bodies.

In the light of the ambiguous power that this image continues to generate, *Patriotismo* in general, and its Fijian episode in particular, may confirm not only that the "condanna spietata dell' 'esterofilia' non significava necessariamente che Marinetti disprezzasse i frutti della cultura straniera" (Härmänmaa 259) or that, while he was still generally sharing many of fascism's ideological positions, he also had a relatively more tolerant and open-minded attitude towards the foreign and the different. More concretely, it may also subtly reveal the chasm existing between the still-present (utopian/ idealistic, almost naively nostalgic), revolutionary, artistic aspirations of Marinetti's old Sansepolcrista fascist self, and the present evolution (or involution) of fascism's more recent cultural polity which, as the regime's level of tolerance decreased, became always more distant in theory and practice from that outlined at the time of Sansepolcro.

Intentionally or not, this "libro giocondo" (91) may ultimately constitute futurism's artistic response and "amendment" to the official Rocco Code (as a prime example of a "codic[e] vetust[o]" [169]) and commentary on fascism's own excessive interference in the sphere of morals; on the gap perceived between fascist legal theory and cultural policies, on the one hand, and every day practice on the other, in particular the kind of practice that was directly affecting the fate and role of futurism itself in Italian culture in these years (such as, most evidently, the closing of *Artecrazia*, the attack on books etc.).[119]

119 Gentile notes that "The principal ideological motif of the Futurist political party during its life from 1918 to 1920 remained the anarchistic utopia of a New State governed by Futurist artists, where the maximum of individual liberty and cultural cosmopolitanism would be reconciled with an intransigent religion of the nation, with imperialism, and with a cult of violence and the glorification of war as the impetus for a perpetually revolutionary modernity, one that would impede the crystallization

Despite the similarities I previously pointed out with Mussolini's own relativistic statement (the one seen as the ideological basis of Sansepolcrismo), the evident personal closeness between Marinetti and the Duce, and presence of the worst kind of fascist rhetoric in the first, sample entry of the new code that constitutes the last chapter of the book ("1. Il patriota che avendo preso a schiaffi un esterofilo denigratore del proprio paese od essendosi divertito a bastonarlo al punto di trovarselo ad un tratto morto fra i piedi verrà delicatamente punito con una pena minima color di elogio e d'incoraggiamento" [195]), Marinetti's aesthetic solution to the challenge of revolutionizing one's "morale" through a surreally humoristic text that praises relativism, in which "well-meaning, far-sighted criminals" tend to be acquitted rather than punished,[120] and that, paradoxically, indicates that an answer to the problems of finding a "Nuova Morale" and "fusing the Individuo con la Patria divina" (91–2) could be found out of Italy and western civilization altogether, among Fijian cannibals who wish to eat books more than anything else, seems to evoke the arrogant rhetoric of the Duce's statement only to indicate its rigidity and actual lack of any "lungimiranza" (farsightedness). This suggestion may perhaps be strengthened if one observes that Mussolini wished to promote fascism as the unique solution to the problems of the fatherland, rather than advocate the potential benefits and advantages of any kind of relativism. The third section of the Programme of San Sepolcro states that: "The

---

of traditionalist inertia and make Italians live fully within the realm of modernity, leading an adventurous, energetic, and quotidianly heroic life." (69). As *Patriotismo* may suggest, this utopian trend does not seem to be absent in futurism's later years although, as I have been arguing, it was combined with a simultaneous sense of an ending.

120 "Considerato che voi deterioratore di don Giovanni volevate evitare il tradimento di mia moglie Paranzella considerato che tu Condirina l'hai avvelenata prima che mi tradisse considerato che voi Pugnante dirigibilista avete sparando Fulgido nella vigna di Lucente Brunswick guarito centinaia di giovani dal vizio dell'adulterio considerato che voi Conte Cremisi dei Vermigli avete bruciato la banca dove avevo intenzione di annegare i miei risparmi io giudice Paranza salvato da voi quindi interessato quindi competente vi assolvo Siano egualemente assolti tutti i soci qui incriminati della Società dei Lungimiranti" (F.T. Marinetti, *Patriotismo* 144).

meeting of 23 March commits all fascists to sabotaging the candidacies of the neutralists of all parties by any means necessary" (O'Brien 13). In other words, Mussolini aimed at promoting the imperialistic, interventist and ultimately dictatorial policies of his own party, and disregarded the notion that different perspectives coming from other parties could be considered at the same time. In this situation, it is not surprising that fascist laws eventually became more authoritarian and oppressive, towards futurism as well.

If, as Tokkamatok says, justice is managed by eating ("mangiando io amministro la giustizia" [184]), and, together with the hydroplane, the food he and his subjects want above all is books ("Liiiiibri da mangiare cogli occhi" [180]), then any diet (such as the current fascist one) that eliminates literary magazines, makes use of censorship, marginalizes books and, by logical extension, the physiological importance and contribution of (futurist) art and culture itself for the healthy body of the nation is necessarily unjust and unsatisfactory. Ultimately, then, the "morale antropofaga" is an aesthethic practice aimed at reigniting the love of one's country. The fusion of the individual and the fatherland can only take place when a constructive moment follows a destructive one or, to put it differently, when one is assimilating, rather than just destructively devouring, the literary "sweet steel" of hydroplanes, and integrating, rather than eliminating, the poetic products coming out of futurist "anthropomorphic" typewriters.

CHAPTER TWO

Modernist Effacements: Aldo Palazzeschi's Consuming Bodies

2.1 Relieving Manoeuvres: Aldo Palazzeschi's *: riflessi*

> An artist is someone who refines himself out of existence.
> — JAMES JOYCE, *Portrait of an Artist as a Young Man*

The chances are that, if asked to name an epistolary novel from the first few decades of the twentieth century, even someone with an above-average knowledge of modern Italian literature would fail to mention Aldo Palazzeschi's 1908 text *: riflessi* (to be read *Due punti riflessi*),[1] also known under the title *Allegoria di novembre*. So far, critical readings of the novel have failed to highlight its epistolary nature and, consequently, they have also left unexplored the interpretive implications this formal characteristic may suggest. They have chosen rather to concentrate on the other prominent characteristics of *: riflessi*, such as the stylistic variants among the differently titled texts, the decadent atmosphere evoked in its pages and, in more

1 The text consulted here is Aldo Palazzeschi, *: riflessi* (Milan: SE, 1990), which includes the helpful postscript by L. De Maria "Il romanzo decadente di Aldo Palazzeschi." The title comes from the poem "Gioco proibito," contained in *Lanterna* (1907) and, as Maria C. Papini observes, "makes explicit the close connection between the two books" [i.e., the collection and the following novel]. See Maria C. Papini, "Aldo Palazzeschi, Remy de Gourmont: Un gioco di *: riflessi*" in *Rassegna della letteratura italiana*, 101.1 (January–April 1997), 116–30, (123).

recent years, its homoerotic content and deep psychoanalytical elements.[2] Even Edoardo Saccone's excellent psychoanalytical reading, probably one of the most insightful available and one to which I shall return in the following pages, tangentially registers the ritualistic aspect of the action and its self-discovering, "reflexive" quality,[3] but does not speculate further on the epistolary factor. The extent of Anthony J. Tamburri's observations on this specific feature of the text, instead, is that we are dealing with an "epistolary novel about two men of an unusual friendship" and that the letters are actually "unsent and, therefore, unanswered."[4]

No doubt Palazzeschi's short novel, with its two completely different, stylistically and structurally contrasting sections – the first and longer one consisting of thirty letters from the protagonist, Prince Valentino Kore to his ex-lover, John Mare,[5] and the second representing the comments of a vociferous crowd after the mysterious disappearance of Valentino – disorients its readers and problematizes any rigid categorization.

But it is also true that the question of choosing an example of epistolary novel from the beginning of the twentieth century is a tricky one. As

2 See for example G. Marcon – D. Trento, *Allegorie e varianti in : riflessi' di Aldo Palazzeschi* (Bologna: Centro di documentazione "Il cassero" (1988); J. Soldateschi, "Da *: riflessi* ad *Allegoria di Novembre*," in *Il laboratorio della prosa: Pratesi, Palazzeschi, Cicognani* (Florence: Vallecchi, 1986), 67–173.

3 Edoardo Saccone, "Allegoria di Novembre: la sublimazione imperfetta di Aldo Palazzeschi," *MLN*, 92–1 (1977), 79–116.

4 Anthony J. Tamburri, "Aldo Palazzeschi's *: riflessi*: Other Reflections of Reading and Writing," *Italian Culture*, 10 (1992), 115–30 (121–2).

5 And one should notice additional "riflessi" not only in the semi-mirroring names of "Ko-re" and "Ma-re," or in the intertextual echo that links John Mare to the name of a minor character ("John Mare, the winning jockey") in Rachilde's 1884 novel *Monsieur Vénus*, but also in the two meanings of the Greek word *kore*: "young maiden (girl, young woman, servant)" (*OED*) and, even more appropriate given the context, "the pupil of the eye." Duplicity and sameness (also in relation to the extension and the limits of the self) as well as problems of presence and absence, are constantly alluded to in a novel which questions the very reality of things. Does a "riflesso" exist or it is just an illusion? The art of writing itself, moreover, is "one means of creating the illusion of presence." Linda Kauffman, *Discourses of Desire: Gender, Genre and Epistolary Fictions* (London and Ithaca: Cornell University Press, 1986), 24.

Jean Rousset writes in one of the few books that deal with the morphology and theoretical aspects of such a category, this particular literary form in Europe "au seuil de l'âge modern" had only an "existence d'un peu plus d'un siècle."[6] As Rousset's observes, this transitional genre emerged at the end of the seventeenth century, flourished during the eighteenth with the help of writers such as Rousseau, Goethe, Richardson, and Foscolo, only to mostly vanish around the middle of the nineteenth century, its disappearance arguably coinciding with Balzac's *Préface des Mémoires de deux jeunes mariées* (1842). Rousset continues by saying that the novel in the twentieth century, while sharing at least a common trait with the epistolary form ("il immerge personnages et lecteurs dans un présent en train de se faire et refuse à l'auteur le point de vue panoramique du témoin omniscient" [71]), recurs only in exceptional cases to the epistolary genre ("Mais il ne recourt qu'exceptionnellement à la letter" [71]). The "journal intime" and/or the interior monologue, in fact, take the place of the epistle as they evolved into narrative devices similarly equipped to convey the protagonist's deepest thoughts and to make the reader "sentir les passions" (69–71). Rousset's critical affirmations have a confident, a-problematic clarity that is in tune with the predominantly descriptive and taxonomic quality of his scholarly text. At the same time, they also convey the impression of a progressive and quite rigidly structured vision of literary history, whereby, for example, it seems appropriate to pick a precise year or a particular text as the unequivocal starting point initiating a whole new genre or "ism" or, moving one step further, still to believe in the possibility of a univocal truth to be found and communicated. Rousset's observations, within these limits, provide nonetheless an excellent overview of the tradition of the modern epistolary genre in Europe and, as such, they are extremely useful to framing my discussion.

6 Jean Rousset, *Forme et signification. Essais sur les structures littéraires de Corneille à Claudel* (Paris: Librairie José Corti, 1962), 76. More recently, on the epistolary novel, see Margherita Di Fazio, *La lettera e il romanzo. Esempi di comunicazione epistolare nella narrativa* (Roma: Nuova Arnica Editrice, 1996).

To return to the question that opened this chapter and to its potential answers, we may realize that the exclusion of a twentieth century text such as *: riflessi* from the epistolary genre can be quite justified (and not only because it is not a well-known work). Actually, it may paradoxically indicate a good knowledge of literary history. One may only wonder if even the mention of what Rousset considers the evolution of the genre – the "journale intime" and the novel characterized by the "monologue intérieur" (71) – would cause a similar laconic response in an imaginary interviewee. My guess is that it would. Other writers such as Svevo with his *Coscienza di Zeno* or Pirandello with his "Colloquii coi personaggi" may come to mind, but Palazzeschi, someone who has consistently, joyfully played with theatricality and collective point of views in many of his poems and novels (*: riflessi* included), would no sooner be perceived as an *intimiste* or an experimenter of "stream of consciousness" techniques than he would be as an (even sporadic) epistolary writer.

What I would like to suggest, however, is that *: riflessi*, with its two sections (another manifestation of the double, "reflex" motif)[7] and its shifting narrative perspective, may actually represent a unique example and, thus, also a fitting missing link, in the chronological development of the epistolary form in the Italian (if not also European) literary tradition.[8] Rousset's assertions, according to which such a development corresponds to the evolution "non pas du moins bon au meilleur, mais du simple au

7 On the presence of more "reflexes" within the novel, one should also point out that entire sentences are repeated, with very slight variations, at different points in the text. For example: "uno specchio ovale dalla cornice tutta sudicia e tutta bucherellata dai tarli, molto mi sono guardato in quella luce qua e là scortecciata nell'amalgama interna formante delle lacune irriflessive" (18) and "uno specchio ovale polveroso e scortecciato nell'amalgama interna formante delle lacune irriflessive" (47); or: "Se chiudo gli occhi talora riesco a rivedere le pieghe morbide di quei merletti sostenuti dai cordoni a nappe d'oro" (40) and "l'intreccio soave di merletti sorretti dai cordoni a nappe d'oro cadenti giù sulle pareti" (56).

8 On this point, see also M.C. Papini's previously quoted article. On the notion of self-parody as a way of clearing the decks for innovation, see Shirley Vinall, "Princes and Pierrots: Palazzeschi's Early Writing and Laforgue," *Italian Studies*, LVIII (2003), 104–32, (122–5).

complexe, de la voix soliste aux grandes organisations symphoniques" (76) are quite appropriate in this context. The novel's own structure (which, once again, is mostly constituted of letters) and, in particular, the shift in point of view from the single one of the protagonist, filtered through his letters, to the multiple one of the commenting public, seems perfectly to follow and materialize on the page the transition from "soliste" to "symphonique" indicated by Rousset.

But this "discovery," of course, is not all. To be able to consider Palazzeschi's first book as an epistolary novel in every regard, even though an "extraordinary" one,[9] is not an end in itself. Such a realization constitutes an initial step that may allow me to evoke, together with the phantasmatic presence of Valentino's dead mother, the ghost of another modernist writer, Franz Kafka, who thought of letter writing as a crucial existential activity, rich in theoretical implications.[10] More to the point, the mentioning of Kafka is an invitation to read Palazzeschi's novel in the light of certain patterns of renunciation, self-effacement, and consumption of the subject – patterns which have been associated with the name and work of the Czech writer and which often recur in a European modernist *milieu* in connection with questions of language and literary communication. It seems, in other words, valuable to approach Palazzeschi's *: riflessi* too under the perspective of "literary anorexia" (and its peculiar variations in this new context) in order to demonstrate – as Heywood does in her own discussion of Kafka's *Letter to Felice* and "A Hunger Artist" – that here, too, we are facing an instance of transformation by reduction of the self in order to make a more "purified" art possible.[11]

To come right to the point, one can say that the protagonist of the novel, Valentino Kore, writes himself to death. "A" death may be more

9 The novel was reprinted together with *Il codice di Perelà* and *La Piramide* in a collection entitled *Romanzi straordinari* (Florence: Vallecchi, 1943).

10 "... Tutta l'infelicità della mia vita – e con ciò non voglio lagnarmi, ma soltanto fare una constatazione universalmente istruttiva – proviene, se vogliamo, dalle lettere o dalla possibilità di scrivere lettere," Franz Kafka, in B. Sebaste, *Lettere e filosofia. Poetica dell'epistolarità* (Florence: Alinea Editrice, 1998), 221.

11 Heywood, 77/83.

appropriate, since, as we shall see, the chance of some form of rebirth is not excluded. His letters seem eventually to consume the very body that writes them.[12]

As Saccone noted, the epistles are precisely "riflessi" or mirrors on which the soul of the protagonist is reflected (Saccone 85). Each letter, as a consequence, is also necessarily a self-address, though one of a particular kind.[13] These are letters that rather than involving, as one might expect, a process of self-creation and self-invention through the act of writing, are apparently conducive to a process of self-effacement, to an erasure of the self.[14] An impression that the reader may get at the end of the novel, therefore, is that Valentino – the sender – who has been writing to find himself has, instead, progressively lost himself in the process. One could

12 On this aspect, see Ellmann, 83.

13 In line with A.J. Tamburri's observation (in "Aldo Palazzeschi's *: riflessi*"), I agree that Valentino's letters may actually *only* be a self-address, since his is the unique voice we hear; there are no replies. It is certainly revealing that, contrary to common habits of letter-writing between friends and lovers, he never invites John to write back and, to emphasize this point further, the reader can never positively infer from the content of any of Valentino's letters that John has replied. In theory, we cannot even be completely sure that John Mare receives his correspondence; his letter may *fail* to arrive. And this failure could happen also because we do not have any proof that John is alive. The text may suggest the opposite from its beginning, especially if, within the omnipresent atmosphere of "dissolvimento," and disregarding the unreliable comments about John's whereabouts in the second section (1), one interprets Valentino's statement "È bello, è sublime il nostro ideato e non deve subire nessun cambiamento" (14) as a potential allusion to the project of a double suicide, and (2), one underlines the fact that Valentino's first perception of John sounds peculiarly "funereal," as it resembles that of an inanimate object (a statue but, also, a corpse) "ti vidi la prima volta divino di freddezza come il più bel marmo" and then "sii sempre così" (14/16). On these and other issues, see Gregory Ulmer's review of Derrida's *La Carte Postale*: "The Post-age" in *Diacritics*, 11 (1981), 39–56. Saccone's reading positively adds to these observations when he states that "il dibattito del melanconico [that is, Valentino] diventa [...] dibattito con se stesso, non, come nel lutto, con l'oggetto esteriore" (94).

14 See Kauffmann, 25.

perhaps stretch this interpretation even further and argue that, to a certain (metaphorical) extent, Valentino himself *is* the letter.

The fact that sending the letters happens over a period of thirty days during the month of November, when the dead are celebrated and which is said to be "bello di irreparabile completo dissolvimento" (16) underlines the "dissolution" motif central to the novel. In such an atmosphere, it is only inevitable that Valentino's de-totalizing gesture becomes somehow equivalent to progressively "sending away" a piece of himself, a "self-mutilating" process that unavoidably leads to his final disappearance.

The interaction between identity and knowledge, as well as the connection between identity and epistolary exchange that Derrida individuates in "Envois," one of the sections of his *La Carte Postale*, may make such a statement less radical than it at first sounds. This may happen not only because Derrida's text is itself partly structured in the form of a one-sided correspondence, or because the letter is raised to the level of a philosophical genre, but also because ideas of obliteration and erasure are here (as, it seems, in Palazzeschi's novel) intimately linked to the true essence of the postal (deconstructively, letters are interesting when they "fail to arrive"). As Gregory Ulmer puts it in his useful reading of Derrida's text: "The entire history of the postal *tekhnè* rivets 'destination' (and 'destiny,' *Geschick*) to identity. 'To arrive ['arriver' – to succeed, to happen] is for a subject to attain ['arriver'] my self. Identity, in all its aspects (truth and being) is the ideology of the postal principle" and "the end of a postal epoch is without doubt the end of literature" (Ulmer 41–2). Further on in his review, Ulmer speaks of "autothanatography" to describe "the death drive upon which Freud speculates in *Beyond the Pleasure Principle*" and the way it "relates to the postal principle [because], like the return inquiry of the postal relay, [...] [it] is the tendency to return to an earlier state" (50). The term does indeed seem to reflect the core situation of a novel in which the protagonist not only undergoes a process of "infantile regression" (literally, a return to an earlier state), but also in which his epistolary writing ends with his virtual, progressive erasure from the page.

To these preliminary observations I may add that both Valentino's persona and his vicissitudes are remembered (and thus re-constructed) only by a series of fundamentally empty "voci." If, in the first section, Valentino withdraws from the world into his "Villa Bemualda" and a virtual silence, one interrupted only by the gesture of writing, by the scratching of the pen onto the paper, so that it may actually happen – as it should in any epistolary situation – that the written word "elude e cancella la parola parlata" (Di Fazio 44), in the second part such a dynamic is overturned. As manifested through the reported voices of various sources (the police, the press, and the narrator), the spoken word takes revenge by first overcoming and then contributing to marginalize and erase with its unreliable, often contradictory and confused noise ("con parole un pochino astruse ma molto notevoli" [130]), the written word and with it, paradoxically, the letters and their author.

As the spoken word takes the place of the written word in the letters in the second part, the narration refers internally to a past event of cancellation, such as Valentino's disappearance, and – in a self-referential and counterintuitive fashion – to the potentially ephemeral (rather than permanent, eternal) quality of the act of writing; to its volatility, rather than its physical persistence over time. An even greater danger than the vanishing of the author and subject of the (literary) text(s) (the "soggetto della scrittura" [Di Fazio, 79]), seems that the letters themselves, intended precisely as "l'aprirsi e il manifestarsi" (Di Fazio 79) of that same subject, and as emblems of (literary) communication, may carry within themselves the signs of their eventual cancellation. "Il richiamo all'allegoria nel nuovo titolo avrà forse più a che fare [...] con 'la distruzione dell'organico e del vivente' propria dell'intenzione allegorica secondo Walter Benjamin" and then "parlare di allegoria [...] vuol dire parlare di scrittura, o, se si vuole, di linguaggio," writes Saccone (87–8). The implications of the novel's epistolary structure seem further to corroborate such conclusions. If Valentino states "E non è per la morte che si vive?" (16) and, as De Maria already noted, in the end he writes to forget, "Pensavo che ti avrei intero ricostruito per dimenticarti intero, che avrei financo contato ad uno ad uno i tuoi capelli per dimenticare ogni loro riflesso d'oro, che avrei rievocato ogni tuo sorriso di sguardi per dimenticare i tuoi occhi ..." (71–2),[15] then Palazzeschi's mission is exactly the opposite. For the novel deals with the attempt to exorcise the deepest fears faced by every writer: a white, blank page, and being forgotten.

The peculiar relationship established in the text between the gestures of writing and speaking and, at the same time, between the issues of presence and absence, asks for one more consideration: Valentino is evidently not the only – or the first – to vanish. As a matter of fact, he is writing precisely because the "other," his "particular friend" John Mare, is absent. The latter has already vanished (he went to Venice) and is physically *not* there. For both Valentino and us, the community of readers, John lives and exists only as the addressee, as he – in a way similar to the mail-addicted protagonist in Giovanni Papini's short story *Chi sei* – (hypothetically) receives and then opens his letters every morning.[16] As mentioned earlier, John is invoked in the narration and, therefore, "is addressed to" for no other apparent reason than successively being forgotten or "rimosso" by Valentino. Valentino's withdrawal in Bemualda and, especially, his unilateral correspondence, thus acquires a very peculiar quality. His correspondence is not, as one may traditionally expect in epistolary exchanges, intended to re-create and re-evoke the addressee's presence as it is, rather, to escape from and elude it once and for all.

Forgetting and erasing are therefore the apparent objectives of the writing process. If this is true, another indirect consequence of this situation is that the parallel gestures of "sending away" a piece of oneself, as the

15 Also in the light of later considerations, it is useful to note that for Simone Weil "to starve is to renounce the past" (qtd. in Ellmann, *Hunger Artists*, 88).

16 Within the context of the link between epistolarity and identity I am (partially) trying to explore, Giovanni Papini's short story "Chi sei?" must necessarily refer to Palazzeschi's poem "Chi sono." See Giovanni Papini, "Chi sei?" in *Il tragico quotidiano* (Florence: Vallecchi, 1906). As J.L. Borges put it, the short story "riferisce la scoperta atroce che non siamo nessuno, fuori dalle nostre circostanze e dalla certezza illusoria che ci danno gli altri, che anch'essi sono nessuno," in Giovanni Papini, *Lo specchio che fugge* (Milan: Mondadori, 1990), 7.

protagonist metaphorically does through his epistles, and that of forgetting, are actually closer than they may seem.

From a theoretical point of view, these acts could both be interpreted as figures of reduction of the self. They are both instances of what could be defined as a "relieving manoeuvre," a kind of lightening of the soul and of the mind of Valentino. As such, sending and forgetting are not only the essential means for Valentino to embark on his narcissistic psychological journey – a journey, we could anticipate, from existential gravity to light – but they also make the protagonist of *: riflessi* one of the earliest embodiments of Palazzeschi's philosophy of "leggerezza."[17]

Going over the ramifications of such a journey (or "corsa," in the words of Valentino), means stressing again the fundamental importance of the initial gesture performed by the prince to the rest of my discussion: his withdrawal or, as he calls it, his "carcerazione" (89) in Villa Bemualda.

Such a move is reminiscent of other memorable situations of imprisonment within (and, of course, beyond) Palazzeschi's work, such as the arsonist in his cage, Perelà in his cell, or Boccanera in his circle, and it is also linked to the entire epistolary tradition.[18] As Ellmann puts it while discussing Richardson's *Clarissa*, a novel where the "misadventures of the post [...] mirror the misfortune of the heroine" (Ellmann 75), "the whole epistolary system of the text depends on her encryptment" (Ellmann 97).[19] It does not take much effort to notice the relevance of her statements to *: riflessi*, in which Valentino, in order to write, needs first to incarcerate himself in the timeless stillness of his silent mansion:

17 On this aspect, see also Saccone, "Allegoria di novembre," 110.

18 As we shall see more in depth in the next section of this chapter, Kafka's *Hungerkünstler* (1922) comes also to mind. In N.N. Glatzer, ed., *The Complete Stories* (New York: Schocken Books, 1971), 268–77. Aldo Palazzeschi, *Il codice di Perelà* (Milan: SE, 1991). "Boccanera" can be found in Palazzeschi's collection *Difetti* (in *Tutte le poesie*. Milan: Mondadori, 2002).

19 Clarissa's situation, in turn, is strongly reminiscent of the one the character of Ottilia experiences in Goethe's *Wahlverwandtschaften*. Ottilia withdraws from the world into a castle and progressively withers, refusing first to speak and then to eat. See Guido Morpurgo Tagliabue, *Goethe e il romanzo* (Turin: Einaudi, 1991), 65–117.

> Sono da forse tre ore chiuso quassù [...] Sono talmente stonato ... sbattuto ... non so orientarmi mica ancora. [...] Eccomi in questa stanza, non ne sento che l'alito della vecchia polvere, non altro, e l'alito mio quasi taccio a questi poveri tarli che imagino ormai padroni, comodi padroni, della villa (13).

The description of Valentino's self-perception here is revealing. It sets the claustrophobic scenario that allows writing to take place and, while doing so, it subtly corroborates the metaphorical overlapping of Valentino with the letters he will be writing. He is literally "penned in," tossed here and there ("sbattuto"), initially without the possibility of getting oriented. As such, he is objectified, reduced almost to the status of a sheet of paper sealed inside an envelope and posted into a mailbox.[20] His self, too – I can now, more confidently re-affirm – is "embodied in the epistles themselves" (Ellmann 98).

More important to my discussion, however, is the realization that his secluded situation alludes to a general pattern of renunciation within the text, one which, by reaching its apex in coincidence with Valentino's final withdrawal from the pages, is undoubtedly a crucial key to interpretation. Because of the range of its implications, the concept of renunciation seems useful in investigating the potential complicities between the acts of writing, self-imprisonment, and self-effacement.[21]

Saccone's interpretation pointed out that Valentino's regression to the "luogo della madre," his psychological quest, can be successful only after he has "rinunciato all'assolutezza del principio di piacere" or, in other words, after he has given up "la vita 'rossa' del desiderio" and substituted it with the [vita] "'bianca' dell'amore" (106–7). Building on Saccone's insights, I would like to argue that in *: riflessi* renunciation and (self-) reduction are intimately related concepts.

20 This lack of orientation reminds, in turn, of Perelà's situation: "mi svoltolavo da ogni parte come un globo di una pietanza indigeribile in uno stomaco umano" (27).

21 For an in-depth analysis of the concept of "renunciation" intended as *Entsagung* (both "sacrifice, suffering, abnegation" but also "self-limitation and self-determination"), see Morpurgo Tagliabue's book.

If we envision the text in its essentiality, what is left until Valentino's disappearance – aptly, since we are dealing with an epistolary context – is a "confessional" frame in the most basic sense of the adjective. The protagonist must confess before receiving "il Disco di Purità" during the ritual of Holy Communion at the end of the first section. Writing his letters evidently fulfills the requirement that a religious confession implies. He needs to purify himself ("mi sono sentito puro in quell'istante primo della mia vita" [104]). He must get rid of (or, if you prefer, forget) and renounce several psychologically heavy objects – (whatever those may be: the ghost of his mother, his "red" life, his love as lustful passion as represented by Johnny) – in order to finally be able (and ready) to reach a higher level of (non-) being, which happens to coincide with the incorporation of the body (and perhaps the word) of Christ:

> mi sentivo tutto riscaldare da quell'aria temperata [...] Le mie ginocchia si sono piegate, ed io sono caduto giù sullo scalino colla faccia sopra la tovaglia [...] ò veduto alzarmisi davanti quel Disco di purità, colla stessa naturalezza del sole stamane: sono dipoi rimasto lungamente con la testa fra le mani (104).

This, we now discover, is the kind of "morte" Valentino was referring to in his first letter to Johnny ("non è per la morte che si vive?" [16]). Whatever this final event represents from a psychoanalytical perspective, the one thing that should especially interest us here is that such a "death" shares the qualities of an ecstatic, mystical experience, and that it happens at the culmination of the process Valentino initiated by isolating himself in Bemualda. Such a process, both for its characteristics and for its final outcome, may legitimately be defined as pertaining to the ascetic sphere.[22]

22 I would also like to emphasize that the final scene of Valentino's communion should not be read just as a strategically placed sprinkling of the "misticismo e religiosità decadenti" (Luciano De Maria, in Aldo Palazzeschi, *: riflessi*, 137) one may find in several novels of these years (from D'Annunzio on) but, rather, as deeply consistent with the general, "ascetic" quality of Palazzeschi's text. For an exemplary study of the relationships between a literary and a theological modernism and the individuation of the link between the "ascetic" and the "aesthetic," see Paolo Valesio, "Spiritualità alla svolta del secolo: un anti-apologo," *Annali di storia dell'esegesi*, 10/1 (1993), 145–63.

"Asceticism," according to the *OED*, means "rigorous self-discipline, severe abstinence, austerity." The correspondent adjective ("ascetic") not only refers to Valentino's overall situation but, in particular, its introduction allows us finally to locate, both etymologically and theoretically, the natural meeting point of the patterns of renunciation and (self)-reduction mentioned earlier. If "self-discipline" is found in the protagonist's voluntary incarceration and in his month of letter-writing-posting rituals, and if "austerity" well reflects the semi-monastic simplicity and lack of luxury of his cell in Bemualda (the "stanza calcinata" [34]), the definition of "severe abstinence" is the one that should especially draw our attention. I assign the notion of abstinence such a value first because, as Kafka's omnipresent ghost keeps reminding us, any act of writing is already a form of abstinence;[23] second, because the notion of abstinence is central both to the "self-consuming process of the modernist writer [and text]" and to the relation of certain "figures of anorexia" with questions of literary creation and communication.[24]

There is no doubt that Valentino is abstaining as much as possible from human contact, from anything that would undermine his effort toward continence and the privacy of his writing. His gesture is a private one and not (yet) in the public eye. His retreat in the villa, then, reveals his reluctance to participate and be engaged in the more social aspects of life and its rituals. His voluntary "carcerazione" betrays his need – from a perspective possibly echoing that of a classic anorexic – to have a routine and some minimal control over his life and, at the same time, of his deep fear of losing both.

No hermeneutical acrobatics are needed to realize that a fundamental anorexic attitude is recognizable in the text. The more the protagonist becomes involved in his letter-writing and his inner search, the less he seems to need to eat, as if – in circumstances, once again, reminiscent

23 "When it became clear in my organism that writing was the most productive direction for my being to take, everything rushed in that direction and left empty all those abilities which were directed towards the joys of sex, eating, drinking, philosophical reflection and above all music. I dieted in all these directions" (F. Kafka, *Letters to Felice*, January 3rd 1912), qtd. in Heywood, 71.

24 See Anderson 35; Heywood 71–84.

of *Clarissa* – the "expression of the word requires the repression of the flesh" (Ellmann 47).[25] After initially appearing to enjoy the ritual "pranzo" prepared by the decrepitly old "custodi" of Villa Bemualda, Cammilla and Imperia, "Piccole cose sane e semplici che forse farebbero campare all'infinito anche me" (20), Valentino shows clear and progressive signs of lack of appetite. In the letter of 10 November, he tells John that "quando la vecchia è venuta per sgombrare la tavola [...] à trovato la mensa ancora intatta." (48); in that of 14 November, he confesses to him that "Non ò potuto mangiare" (65); on the 17th, that he is "assai assai malato" and that he did not want to "prender cibo" (74); on the 23rd, just a week before his actual vanishing, the symptoms reach their culmination, as he writes "Da tre giorni non ò preso cibo, e sono sfinito, impotente di cercare e di pensare" (87). This pattern towards a potential "dissolvimento" (16) of the subject by starvation, however, is neither linear nor consistent, and it is actually interrupted after his successful "return" to the mother. Following this event he feels "guarito," very hungry and ready to binge: "mi sono di subito seduto a mangiare presto presto con un'avidità straordinaria, tanto che ò in breve finito tutto tutto, e certo che se non me ne fosse mancato il coraggio io avrei domandato ancora qualcosa per il grande appetito che avevo" (91). While the recurrence of Valentino's detailed comments about his food intake throughout the novel is intriguing, their meaning may become clearer if one adds that these references relating to the semantic field of food are not limited to the description of his own eating habits.

The loathed "piccola bettola" (19) he must walk by every time he has to mail one of his letters – a public place where people gather and speak loudly while eating and drinking – becomes an embodiment of (one of) the protagonist's quintessential fears, as a description such as "lo spiraglio rosso d'inferno" (23) implies. That such a "bottega" (25), on at least two further occasions is directly compared to a mouth ("non ò potuto vedere chiusa la piccola bocca dall'alito pestifero" [47]; "la piccola bottega rigurgiterà i suoi lazzi scoppiettanti e striscianti laidamente" [54]), and

25 We should also note that his behavior is consistent with the convention of fasting before the Communion ritual.

repeatedly threatens Valentino as he comes near it ("ò tremato, ò avuto paura nell'avvicinarmi" [19]; "è certo che le fiammate pestifere verrebbero fino ad attossicarmi," [54]), seem to further suggest that the fear we are dealing with here is a kind of "oral fear."[26] This oral motif is subtly introduced as early as the first page of the book. My allusion is obviously to the description of the "poveri tarli [...] comodi padroni della villa" (13) that are slowly eroding the very structure of the protagonist's abode with the mere power of their mandibles, in what seems an early, eerily accurate projection of what will become one of Valentino's deepest obsessions: that of being eaten away. In other words, if anorexia in its original meaning can loosely be defined as a fear of eating, one of its peculiar variations in *: riflessi* seems to be rather, and more appropriately, a fear of being eaten. The progressive overcoming of the literal and allegorical implications of this fear will be the crucial objective of Valentino's "corsa."

The observation that the "tre enormi pagliai colla loro grande pancia banale" (30), "ripugnante" (34), "sozza" (55), "enorm[e]" (58) – Valentino' s "incubo giallo" (34) and evident emblems of his oral fear, of the dangers of being swallowed up – are not only anthropomorphized, but are described as a kind of communicative act, which is perceived in extremely negative and hostile terms by an intimidated Valentino, may let us catch a glimpse of the finishing line of such a "corsa" as well.

When the "pagliai" first appear, Valentino observes that they are "spavaldi per la loro vacuità" (30), and in this statement "vacuità" can pertain to a structural but also to a symbolic sphere (they have an empty shell but they also emit a vacuous discourse). Further on, and more to the point, he worries that they constantly conjure and gossip against him with "dei loro discorsi inconcepibili [...] i loro discorsi impunibilmente colpevoli"

26 It is difficult here to dismiss an inter-textual echo with the episode of Huysmans' *À rebours* in which des Esseintes enters a "Bodega." As Carol A. Mossman notes: "nothing is more physical than des Esseintes's visit to the Bodega which takes on all the allegorical flavor of a descent into the human belly. The pub into which des Esseintes slips is described as a 'panse garnie de créneaux de bois,'" in "Gastro-Exorcism: J.K. Huysmans and the Anatomy of Conversion," *Romanic Review*, 79 (1988), 341–53 (346).

(34). He even imagines having a conversation with them: "ò parlato, ò rivolte alcune parole ai pagliai in risposta ad alcune loro. Stretti nei loro conversari, inintendibili essi non m'hanno risposto affatto [...] si sono appena sussurrata qualche parola tronca con la massima indifferenza" (55).

Moreover, right before he burns them down with his cathartic gesture (when the power relation begins to shift), the "pagliai" are said to still smile impassively but also to "non [avere] ora cuore di pronunziare" (58). Picturing these three haystacks as the obese silhouettes of traditional men of letters, possibly sitting outside the Florentine "Caffé delle Giubbe Rosse" (a "bettola" of its own special kind?) as a young author tries to break into the literary scene with his first novel, is perhaps too daring a proposition and one that would erroneously make the protagonist of the novel coincide with its author. However, if we keep in mind that "Palazzeschi's *ars poetica* is founded on his refusal to accept the cultural heritage at the beginning of the century" and that "his only inheritance from the past is the 'spazzatura/ delle altre poesie,' as he ultimately reduces poetry to *garbage*" (Tamburri 1990, 117), such an outrageous hypothesis may give a hint of what potentially lies beyond the allegorical veil of the text.[27]

Together, these observations lead to the suggestion that the principal effect of the first fire encountered in the novel is to silence one kind of discourse – one which is vacuous and cannot be comprehended – and, at the same time, to create the premises for the emergence of another, alternative one: his own. As the bonfire puts an end to an established dynamic of intimidation and reciprocal incomprehension, it also sanctions the beginning of the end of Valentino's own oral fear (even though he does not realize it at this time). That the accomplishment of such an event is significantly marked by an ingestion performed by Valentino should confirm both his novel active disposition (i.e., he "swallows" and is no longer fearful of being swallowed) and, by the same token, reiterate the relation

27 Tamburri's observation refers to the poetics Palazzeschi adopted in his futurist phase. However, since Palazzeschi's refusal of a previous poetic tradition and his spirit of innovation do not seem to be limited only to such a period, I am using it in a more comprehensive fashion.

between language and food to which I have been referring. In a tangential but significant comment, made immediately after he has performed his arsonist's act, he realizes with some amazement that he is now able to "ingollare un riso convulso rumoroso doloroso che [gl'] invase la gola" (59).[28] I read this act as the first sign that Valentino's journey has finally found its direction. It is easy to see that such an act is anticipating, not to mention mirroring, the other gesture of ingestion (the Host) that takes place just before he sets his last, all-consuming fire, the one that will enable him to get involved in a quite different kind of consummation-communication.

The physical elimination of the menacing "pagliai," in this perspective, is equivalent to the eventual closure of the "bottega." This is the other object which threatened to eat him and which – now we realize it even more clearly – was also regurgitating a despised communicative act from its mouth, with its "lazzi scoppiettanti e striscianti laidamente" which, under their shape of "fiammate pestifere" could even poison him (i.e., divert him from finding his own voice): "La bettola, la piccola bottega era chiusa! [...] Tu comprendi vero come tutto questo abbia risvegliato il mio spirito ..." (101). It is only when the "bottega" has stopped the emission of its own pestilential discourse that Valentino may state that his spirit has been reawakened and that he is now "smanioso di agire" (100), finally on the right track to both get rid of his earthly weight and to find the way to his own production of meaning.

The many references to water (rain) and mud in the letters need also to be briefly addressed in order to complete my reading in the perspective I have delineated. The presence of such elemental matters seems to belong precisely to this earthly and heavy, corporeal reality in the narration which, eventually, needs to be left behind. Valentino's journey, consistent with this line of interpretation, could very well be a passage from the difficulties

28 In the context of asceticism and mysticism, one may note that, in tune with the literary tradition which attempts to represent such experiences, the fire has physical effects on Valentino's body: "la mia temperatura dové cambiare di almeno tre gradi, non so se io divenni di fuoco o di ghiaccio" (59).

and dangers of writing and communicating to the discovery of one's own higher and better voice.

In order to send his letters, Valentino must not only overcome the engulfing fear of the "bottega" and that of the despising "pagliai" with their protruding bellies, but he also needs to "tuffar[si] nella melma, imbever[si] d'acqua" (4 November [25]). In the following letter (5 November), he seems to indicate that such a low, passive, sinking condition, be it either "nel fango" ("E come saprò muovermi stasera affondando nel fango" [28]) or, similarly, "affondato nella poltrona," is eventually bad for communication ("le mie labbra non potranno più dire" [28]). Further on (in his starving phase), he writes that "il mio mantello è a terra sola fanghiglia e le sue pieghe non sono che cavi di fango che cedono all'insinuazione dell'acqua melmosa, e il mio corpo ne sale fuori viscido, cascante di gialli untuosi che vi serpono schifosamente [...]. Come [...] un rospo che si contorce nel fango senza poterne uscire" (87). Even more revealingly, he then asks John to both forget and remember him ("fa di obliarmi e di ricordarmi") "come una sozzura della quale tu ti sia completamente liberato" (96). If expanded beyond its most immediate context, this request to John expressed through the scatological reference betrays a fundamental wish Valentino unconsciously has, one towards which he has been striving throughout the narration (and which, as we shall see, recurs under a different shape also in *Il codice di Perelà*). To ask to be compared to (and remembered as) an expelled "sozzura," one which has gone through its path – (in other words, not as something like a toad stuck in the mud, but as something/one which has seen the light, which has been moving from inside to outside) – may not simply constitute one of the "attacchi sadico-anali" Saccone talks about in his discussion (Saccone 101).[29] Especially if one sees these lines as a not-so-veiled reference to what Ellmann calls the "excremental origin of speech" (Ellmann 47), this comparison also implies that Valentino's goal is to be remembered as someone able to create his own "discorso." Such a

29 If a sadist is (also) a subject who constantly tries to establish a condition of "togetherness," only to be able to affirm later his separation from the other/object, Valentino may well belong to this category.

circumstance may take place only once he has managed to overcome the devouring threat of the "bottega" and the "pagliai" (that is, the fear of being eaten [by] their "discorsi"), and he has emerged from those dark, muddy, intestinal lowlands which have been making him sink/sick and which have made communication (and enlightment) difficult – if not impossible – for him. To desire association with an expelled "sozzura" which has come to light (leaving aside all the biographical and sexual implications which are certainly involved), means implicitly that the fear of being eaten is a thing of the past (because one has already gone through the tunnel). The loss of oral fear, in turn, is directly linked to the possibility of finally producing and emitting one's own creative discourse.

The ultimate overcoming of Valentino's oral fear, however, cannot come from a passive act (being expelled), but from an active one. This should explain why the ingestion of the Host comes into play and sanctions, in coincidence with the second, final fire, Valentino's eventual finding of a truly alternative word. Consistent with the landscape of this text, one which deals with yet another variation of modernist anorexia, his final ingestion significantly becomes the "unica consumazione possibile" (Saccone 111). This event is one that, as the word "consumazione" itself suggests, immediately pre-announces the following, literal consumption of the protagonist, his material disappearance amidst the purifying, ever-transforming fire. Similar to Boccanera's fire in the homonymous poem by Palazzeschi, the Host is the only kind of consummation with the effect of *not* increasing the weight of the subject who is ingesting it and which, on the contrary, en-lightens him.[30] Or rather, eating the Host both illuminates the subject and makes him lighter. Valentino's observations, as the moment of the ritual approaches, contribute to describe a general atmosphere of increasing light and lightness. The exterior reflects and blurs with the interior as his com-

30 The Host is also the sort of food that "converts the feeder into himself [...] The act [of eating the Host] is one of reciprocal incorporation, as both are identified by the single word and substance, the Host, so that the absolute boundary between inside and outside, eater and eaten, itself appears to disappear." Maggie Kilgour, *From Communion to Cannibalism. An Anatomy of Metaphors of Incorporation* (Princeton: Princeton University Press, 1990), 15.

ments on the weather echo his own status, until "la pace" fills him in: "il nero, l'oscuro, si è poco a poco venuto scaricando [...] il grigio si scaricava, si alleggeriva"; "la mia testa non pesava punto" (102–4).

Once again, however, this kind of self-consumption (or "sublimazione," to use Saccone's term) needs to be interpreted as the necessary premise for a re-birth, an aesthetic rebirth to be precise, and one which may make the production of a new, different kind of (literary) discourse possible. The conclusive passage in Valentino's last letter seems further to reinforce this interpretation. If my previous observations are legitimate, his final descent, opening the doors, and then exiting the fiery bowels of Villa Bemualda (leaving behind the "peso di questa stanza calcinata") in order to send out his last message ("i merletti leggeri, finissimi" [56]) while feeling the closeness of (day)light, not only mirrors the enlightening, descending path of the "sozzura," but also hints at a precise kind of artistic operation:

> Comincerò ad aprire la mia finestra e scenderò rapidamente, poi per uscire ad impostare non passerò dalla piccola porta segreta ma dal portone grande, e scenderò lo scalone per la prima volta in questo tempo.
>
> Io stesso aprirò il portone in segno del principio della mia festa.
>
> Deve essere tardissimo, mi sembra che sia notte da tante e tante ore e che il giorno debba essere vicino. Mi vedessi Johnny, sono tranquillo e sereno, e bello, tanto bello stasera!
>
> Addio (108).

Anticipating Perelà's parallel dropping to light from a chimney and the divulgation of his philosophy of "leggerezza," Valentino (and, by extension, Aldo, the writer) is telling us that from now on he will take one path rather than another in order to create and communicate (as the act of writing and the following one of "impostare" suggest). As he seals his last epistle, he opens the door, finally allowing his party to begin and sanctioning his entrance into the world of aesthetics not (passively) from the "piccola porta segreta" (anal), but (actively) from the "portone grande" (mouth).

That Valentino's "corsa" ends with the affirmation of his (apparent) narcissistically proclaimed beauty, that is, only when a quintessentially aesthetic objective has been achieved ("sono [...] bello, tanto bello stasera"), should both corroborate the allegorical nature of Palazzeschi's text and help to reflect on the peculiar nature of this proclaimed beauty and its existential implications.

As Paolo Valesio reminds us, "la bellezza richiede, per essere mantenuta e valorizzata, un costante e difficile esercizio, dunque un' *ascesi* nel senso etimologico del termine" (Valesio, 162). Valentino's overall disposition and final disappearance seem to suggest that the achievement of this kind of beauty is associated with an act of ascetic renunciation, with a sacrifice and an erasure of the self.

The final "riflesso" of the story, therefore, is the paradox that language has effaced the self in the very process of attempting to re-constitute it. It is especially because of such an event (an anorexic one), that a renewed and/or different (aesth-)ethical presence may take shape and keep finding its way in the world.

2.2 Vanishing Acts: Kafka's *A Hunger Artist* and Palazzeschi's *Il codice di Perelà*

> What is most surprising is rather the body; one never ceases to be amazed at the idea that the human body has become possible.
>
> — F. NIETZSCHE

In the previous section, the brief appearance of the ghost of Franz Kafka helped to individuate a pattern of renunciation and self-effacement in Palazzeschi's *: riflessi*. My objective in the following pages is to discuss in greater detail a few instances of this pattern as they manifest themselves in some other works by these two modernist writers.

A comparative analysis of a few representative texts under the perspective of the "disappearing subject" – Kafka's "A Hunger Artist" (1922), several of Palazzeschi's earliest poems in the collection *L'incendiario* (1910), his poem "Boccanera" (1915), and especially his novel *Il codice di Perelà* (1910) – not only confirms these authors' new and creative solutions to the question of the representation of the artist and the poetic self in a European modernist *milieu*, but also highlights some connections between two figures who, at first glance, could not appear more dissimilar.[31] I hope to identify those common aesthetic principles that may have led them to understand, interpret, and resolve some fundamental artistic and poetic questions in a similar manner.[32] What surfaces in their works when read side by side is a series of peculiar resonances and analogies that, in some cases, demonstrate "adjacent" (to use Deleuze and Guattari's term) cultural attitudes or inclinations and, in others, perhaps reveal only similar flights of imagination. A statement by Tamburri supports my point:

> discussing Palazzeschi and other Italian and Western [and, by the same logic, also Eastern] European artists does not necessarily imply a direct, and conscious, intertextual relationship. Instead, the premise here is one of an epistemological and, perhaps also, ideological commonality – namely, they were all artists coming into their own during an aesthetically turbulent time (Tamburri 1998, 5).

It may be useful to linger for a moment on this era. The situation of the artist in this turbulent period may perhaps be best defined, in aesthetic as well as sociological terms, by the paradoxically similar conditions of either marginality or privilege. Marginality occurred when the artistic product was a commodity, dependent on the market, on the whims of the masses, or when it remained confined to limited, private circles of *connoisseurs*. Privilege, conversely, may have depended on the popular belief that art

31 N. Perella and R. Stefanini, in their Introduction to the English translation of *Il codice di Perelà* (*Man of Smoke*. New York: Italica Press, 1992) already observed that "a parallel can be drawn between *Man of Smoke* and Franz Kafka's novels," although, for obvious constraints, they did not elaborate on their intuition.

32 As a matter of fact, Kafka's *oeuvre* is absent from Palazzeschi's otherwise very well furnished library: "La biblioteca comprende anche una vasta scelta di opere di autori stranieri. Tra queste risaltano i romanzi e i racconti di Dostojevskij, Tolstoj, Gogol, Cècov, Puskin, Turgenev; quelli di Zola, Gautier, Balzac, Flaubert, Maupassant, Stendhal; le poesie di Laforgue, Baudelaire, Rimbaud, Rodenbach, Verlaine, Valéry, Rilke; gli scritti di Goethe e l'edizione delle opere complete di Nietzsche." (Qtd. from the official site of the "Centro Palazzeschi": http://www.palazzeschi.unifi.it/mdswitch.html [Last accessed November 2010].)

was the ultimate repository of meaning, of truth, and that the artist was the only one able to oppose effectively the spreading commodification and vulgarization of society. Within this context, the act of "disappearing" or "withdrawing" from the text becomes a gesture that is at once a reaction to a negatively perceived status quo and a signal of the crisis of the ontological and cognitive function of language.

In other words, both Kafka and Palazzeschi were likely attracted to the formal trope of a "self-effacing" subjectivity because they shared a Nietzschian critical attitude towards history and, in turn, towards a reason/rationality that historically has been located in the subject.[33] As Nietzsche reminds us, building on a tradition of thought which moves from Rousseau to Freud, the cult of history that afflicts modern man is a paralyzing one, as it prevents any active and creative disposition in the present as well as in the future. In an original and radical interpretation of Nietzsche's insights, both artists may allegorically turn to the ultimate gesture of disappearing as a way to undermine the notion of a "subject" who is the "fondamento della ragione occidentale" (Moroni 21) and to escape from the heavy "cage" and the "sickness" of history.[34] In this light, the self-effacement mode becomes the self-ironical and paradoxical solution to "end" history and, at the same time, free oneself from its stifling influence. Such a liberating gesture, is not, however, an exclusively nihilistic one, since it leaves the space empty and ready for the (utopian) possibility of a new origin or a re-birth. By renouncing possession of the world in the belief that this "sacrificial" act may provide access to a higher truth, what Kafka and Palazzeschi end up sharing is a spiritual attitude, one that merges the notions of the aesthetic and the ethical. That such a renunciation manifests itself through some form of reduction seems logical. After all, as Heywood observes, the purest

33 Palazzeschi and Kafka may well have been influenced by Nietzsche's thought especially as it is manifested in his *Untimely Meditation, On the Uses and Disadvantages of History for Life* (1874) and *The Birth of Tragedy* (1871). On the relationship between Kafka and the philosopher, see Patrick Bridgwater, *Kafka and Nietzsche* (Bonn: Bouvier, 1974). Piero Pieri in his *Ritratto del saltimbanco da giovane* explores Palazzeschi's own indebtedness with Nietzsche's work.

34 On this, see (Guglielmi 1993, 158).

kind of art can emerge only out of an act of subtraction and elimination of the "excess," be it a spatial or a temporal excess. As the finite, material body "disappears into the infinite text" (Heywood 81), affirming the primacy of the sign over everything else, the "burden" of temporality is also eliminated, thus creating a *tabula rasa* and the (illusory) pre-condition for a different kind of creation.

From this exceptional, eccentric position the artist's communication tends towards either a language of abundance or one of "reduction." Thus, at one extreme we may find a bombastic discourse that seemingly seeks to create an immediate link to its audience, as exemplified by D'Annunzio or Marinetti in the Italian context. At the other we may perceive a subtler, allegorical and "extraordinary" discourse, one shaping what I would call a "subtractive" (though no less powerful) rhetoric. And it is, I believe, this "subtractive" rhetoric, often represented linguistically by silence and visually by an absence or a void, which constitutes another important connection between Kafka and Palazzeschi.

A caged arsonist dangles in front of a jeering and disdainful mob in Palazzeschi's poem "L'incendiario" (1910); a "man of smoke" in *Il Codice di Perelà* who "non dorme, non mangia, non fa nulla" (81) undergoes a trial and then vanishes from his prison cell, leaving in his place a pair of boots; a fire-eater in "Boccanera" progressively dematerializes in front of the reader's eyes together with the "autophagus" poem that contains him; and – in a visual and thematic collage of all these texts – a professional "fasting artist" exhibited by his impresario in a "small barred cage" to a curious crowd who, after "the verdict went out against him" (Kafka 1971, 276), erases himself to death, leaving the stage-cage to a panther. This, of course, happens in Kafka's enigmatic parable "A Hunger Artist" (1922).[35]

35 One could further bolster this comparative list by obviously adding *: riflessi* (1908), the poems "Il principe scomparso" and "Il frate rosso" (in *Poemi*), and also Kafka's last story, *Josephine the Singer or the Mouse Folk* (1924), whose end is marked by another enigmatic disappearance ("That happened a day or two ago; but the latest is that she has disappeared, just at a time when she was supposed to sing"), (Kafka 1971, 375).

What these stunning and unforgettable images do is tempt the reader to consider other analogies that the plots (or anti-plots) of these works may suggest: a shared fascination with the relationship between the one and the many or, from a closer perspective, with the (sadomasochistic) confrontation between a single, somehow "different" creature (the artist/performer as *saltimbanque* and booth-wonder) and an undistinguishable crowd.[36]

These lone disappearing characters share a subversive inclination to question the rules, challenge authority, and/or test the limits of the current law(s). Either in the attempt to "burn the world" and establish a kind of enigmatic new "code" – the tasks, respectively, of the "incendiario" and of Perelà – or in the similarly puzzling desire to break a fasting record, the accepted rules of society, the physical laws of the body, and, last but not least, the literary norms that, until then, had regulated a "normal" text (and "Boccanerá" stands out in this regard), are simultaneously infringed.[37]

36 At the level of superficial analogies, if one dares to gaze beyond the horizon of these works, a common tendency to make an allegorical "use" of animals (just think, for example, at Palazzeschi's collection of short stories entitled *Bestie*) may also be noted. From a biographical-psychoanalytical perspective, moreover, one could mention, for both these writers, the crucial importance of the figure of the father. On this topic, with reference to Palazzeschi, see (Curi 1974), (Pieri 1980). In addition, one may notice the religious undertones in these literary works that the sacrificial acts of burning, eating fire, and fasting call to mind. "Adam is granted the power of naming the creatures of Paradise and he was later expelled for having eaten from the Tree of Knowledge. Mystics fast in order to prepare themselves to receive the word of God. The body of Christ, the word made flesh, is eaten in holy communion" in Paul Auster, *The Art of Hunger* (Los Angeles: Sun & Moon Press, 1992), 33. On the Christological theme in Palazzeschi, see, for example, Luciano De Maria's essay in *Il codice di Perelá* (Palazzeschi 1991, 203–10).

37 There should be no need to repeat well-known statements about the innovative and ground-breaking quality of these authors when compared with a traditional idea of poetry and prose. Palazzeschi's role in the process of renewal of twentieth-century Italian poetry is well known. Regarding Kafka, Michael Bell notes that: "[his] enigmatic simplicity incites interpretation, a need for meaning, only to frustrate it. The anguish of Kafka's fiction, whatever its other causes or implications, comes from a desire still to find, rather than create, a meaning" (qtd. in Levenson 14).

From a merely topological perspective, the most semantically crucial space[s] of these texts (the arsonist's and the hunger artist's cages, Perelà's cell and Zarlino's madhouse in *Il codice di Perelà*, and Boccanera's public stage) underline the characters' separation from a traditional social and cultural context. That these spaces may not only isolate but also protect these characters from the external world should not question the fact that we are presented here with a first visual sign of an epistemological crisis. We are facing, in other words, a series of contexts in which the subject is no longer able to appear integrated and which may subtly announce his future complete disappearance (which is also a sort of "disintegration") from the fictional world that incorporates him (Pieri 124).[38]

Against this background, it may also be rewarding, to a wider discourse concerning the general characteristics of European modernism, to focus on the nihilistic tendencies that characterize the works of these writers. In this sense, comparing Palazzeschi's "seriousness" and "profondità," masked behind his ode to lightheartedness "Lasciatemi divertire!" and Kafka's *jouissance* and irony, even more artfully hidden behind the disquieting scenarios of many of his stories, is particularly productive. As it would be useful, along the same lines, to read the former's philosophy of "controdolore" and "divertimento" vis à vis the *joie de vivre* which, at least according to Deleuze's and Guattari's authoritative reading, represents the latter's true nature.[39]

38 And one can probably recognize a little bit of Zarlino, il "pazzo volontario," in *Il Codice di Perelà*, within the *Hungerkünstler*; he too a sort of "pazzo come [vuole lui]" and someone who pursues his own "performance," according to his own private and mysterious "code."

39 "Only one thing really bothers Kafka and angers him, makes him indignant: when people treat him as a writer of intimacy, finding a refuge in literature, as an author of solitude, of guilt, of an intimate misfortune. [...] There is a Kafka laughter, a very joyous laughter, that people usually understand poorly. [...] He is an author who laughs with a profound joy, a *joie de vivre*, in spite of, or because of, his clownish declarations that he offers like a trap or a circus." Deleuze, Gilles and Guattari, Félix, *Kafka. Toward a Minor Literature*, trans. by D. Polan (Minneapolis-Oxford: University of Minnesota Press, 1986), 41.

A few questions serve to bring us back to the texts. What, after all, is the hunger artist doing through his fasting, if not becoming more and more "incorporeal" and "light" in order to clear the way for a "higher," more "joyous" form of artistic attraction in the shape of the panther?

> his body was hollowed out; his legs [...] scraped on the ground as if it were not really solid ground, as if they were only trying to find solid ground; and the whole weight of his body, a featherweight after all [...] the joy of life streamed with such ardent passion from his [the panther's] throat that for the onlookers it was not easy to stand the shock of it (Kafka 271/277).[40]

In an analogous way, what are Palazzeschi's protagonists doing in "L'incendiario," *Il codice di Perelà* or "Boccanera" if not thinning their bodies to ashes, smoke, and a single F[lame]? Are these bodily remains not strikingly similar to those left by Kafka's starving protagonist? And what are these characters doing with these remains if not symbolically leaving behind an enigmatic "code" for the production of a new kind of artistic (literary) work? "Eccolo, questo solo può essere il Codice ch'io vi lascio, esso custodiva sulla terra la mia sola virtù" (*Il codice* [197]); and "corri, a riscaldare/la gelida carcassa/di questo vecchio mondo!" ("L'incendiario").

Thus it seems that some sort of sacrifice is necessary both to provide closure for these individual literary works and to make possible the realization of any future artistic achievement.[41] Read in this light, the dedication that opens *Il codice di Perelà* becomes particularly emblematic: "Affettuosamente dedico: al pubblico! Quel pubblico che ci ricopre di fischi, di frutti e di verdure, noi lo ricopriremo di deliziose opere d'arte." If one looks beyond Palazzeschi's omnipresent irony, the dedication implies the idea of a sacrifice on the part of this "messianic" writer-martyr, who claims

40 "The hunger artist's self-cancellation has cleared the way for this figure, a necessary stage of diminution, in order to arrive at greater richness and fullness, which for Kafka is only possible – and even then, this possibility is highly qualified – in the realm of art." (Heywood 77).

41 I am thinking, in particular, of Renè Girard's observations on sacrifice as a founding gesture necessary for the birth of civilization (an "artistic civilization" in Perelà's case).

that he himself shall submit to a series of offenses and insults for the higher (artistic) good of an aggressive and ungrateful multitude. But there is more: its references to a gastronomical reality, mentioned in connection with an aesthetic one ("frutti," "verdure," "deliziose opere d'arte"), must necessarily ring a bell in the critical context I am exploring here, which aims to address the aesthetic implications of fasting, weight-loss, and burning, and their role in any process of effacement.

If these considerations are legitimate, these works by Palazzeschi and Kafka are not only extraordinary examples of modernist literary invention and creative skill. In dealing with questions of aesthetics and issues of artistic creation and communication, and addressing the role of artistic subjectivity from eccentric and innovative perspectives, they may be approached once more from the perspective of the "modernist anorexic logic" recently applied to *: riflessi*. That is, in their wealth of meanings, each of these literary works by the two writers present instances of "transformation by reduction" of the self in order to make art (here, specifically, literature) possible.

In other words, using the critical perspective of "literary anorexia," the connections between the central act of fasting (and writing) in Kafka[42] and instances of burning (and eating) in Palazzeschi relate, in one way or another, to the question of self-effacement.[43] This is not only because fasting is itself, in a way, a specific kind of burning (i.e., of calories, fat) or because of the explicit connotations of purification and renewal these acts possess but, once again, because they are both "irrational" and "anti-historical" gestures which are the "necessary precursor[s] for the transmutation of matter into higher forms" (Heywood 79). These acts are attempts to react to the limits and the finiteness of the spatio-temporal coordinates and to reach that "a-topic" dimension where the subject objectifies "il senso dell'esistenza dal punto di vista della sua assenza e *separazione* sociale" (Pieri 124), and where the "material" vacates its place for the "spiritual," where new possibilities and a new kind of artistic communication may eventually

42 "To speak, and above all to write, is to fast." (Deleuze-Guattari, 20).

43 See also (Pieri 121) for a discussion of the image of the "naked man" in an expressionist context.

emerge. Reading Kafka's and Palazzeschi's texts against each other through this kind of "anorexic" lens – one which is "central to canonical modernist ideas about literary creativity" (Heywood 80) – reinforces their belonging to a common European modernist milieu, and provides an interpretive lens with which to clarify the significance of self-effacement with regard to modernist subjectivity.

A poem such as "Boccanera" (meaning literally "black mouth" and originally published in *La Voce*, January 1915) exemplifies some of these issues. Though it has repeatedly attracted the attention of Palazzeschi scholars (see Curi, Lepri, Tamburri and, most recently, Cangiano), it has never been read through this "anorexic lens." "A metaphor for Palazzeschi's poetry" and "the final stage in the metamorphosis of the *homo dolens* to *homo* ludens" (Tamburri 1998, 94–5), "Boccanera" presents us with a pyrotechnic show performed by a single artist in front of an audience and, as in Kafka's story, this performance is literally constructed around the symbolic action of eating:[44] "Ogni sera su quest'ora/mangia il fuoco Boccanera/[...] Tutta la vita à mangiato il fuoco/[...] Mangiò il fuoco anche l'altra sera/[...] Mangerà il fuoco anche stasera certamente". Boccanera's performance, in other words, may be interpreted as an act of "self-effacement" (Anderson 244) where the protagonist contributes both to the existence and the "death" of the text, since he vanishes at its end. The now familiar image of fire dynamically merges with that of eating in a poem whose "exterior" textual development is the logical inverse of the effects that its main "interior" topic (eating) exerts literally and metaphorically on a physical or textual body: that is, we are witness to an act of progressive consumption, rather than to an increase in size, weight or length. "The body of the text progressively undermines and ultimately contradicts the affirmation of its title," writes Anderson about two of Kafka's most baffling short stories, "A Hunger Artist" and "Josephine the Singer" (246). I would argue that a similar and related dynamic is also present in Palazzeschi's "synecdotic" poem, where the "black mouth" of its title – the organ indispensable to

44 The act of eating appears in several other poems by Palazzeschi (i.e., "La morte di Cobò," "Disappetenza," "Il pranzo"), but it is never as central as it is in "Boccanera."

producing language as well as to ingesting food – first undermines and then literally "eats away" at the body of the text.

Fire, after all, is no ordinary nourishment, and its effect on the performer who eats it aptly fits the kind of reversed perspective and deformation of reality that characterizes Palazzeschi's writings from the manifesto *Controdolore* on. Even though he is "eating," there is no augmentation in the fire-eater's physical weight or size, just as there is no increase in the number of words necessary to illustrate the process. Rather, the fire-eater experiences in his persona something closer to the effects of a prolonged fasting analogous to the one carried out by the *Hungerkünstler*: he progressively vanishes from his improvised and circumscribed stage in the street and from the poem, as an approaching street-cart threatens to roll him over:

> Mangiare ... il fuoco [...]
> Le ruote di quel carro! Le ruote di quel carro!
> Oh!
> Uh!
> Fuoco
> Fuo
> Fu
> F

Something is left behind, however. Something that – like Kafka's fiery panther – may well be "leaping around" symbolizing "the joy of life": a single letter that is the shrunken remnant of a former totality but also represents the possibility of a new beginning. A single, "leaping" and "joyful" "F"[lame], after all, is all it takes to rekindle a fire that will be able to "riscaldare /la gelida carcassa/di questo vecchio mondo!" as Palazzeschi reminded us in the last verses of "L'incendiario."

All this suggests that the implications of the fire imagery in Palazzeschi are even richer than previously believed and may spark an association with the recurrent image of the flame in Kafka. For Palazzeschi the fire icon signifies more than "the rejection and destruction of the undesirable – namely, restrictive social norms and conventions" (Tamburri, *Of Saltimbanchi* 151). Both his famous verse in "L'incendiario" – "vorrei scrivere soltanto per bruciare" – and my reading of "Boccanera" suggest that the act of writing in order to (be able) to burn must be complemented by the equally necessary and mirror-like (and eminently Kafkaesque) one of figuratively burning in order to be able to write.[45] "Quando tu bruci/tu non sei più l'uomo,/ il Dio tu sei!"; "E dopo mi sentirò lambire le vesti,/le fiamme arderanno/ sotto la mia casa [...]/griderò, esulterò,/m'avrai data la vita!" exclaims the poet addressing the arsonist, his words perfectly expressing the idea that it is only in the process of burning (and also, one may argue, disappearing) that man can achieve the status of a true "creator" and give birth to a purer art. Burning is for Palazzeschi, as fasting is for Kafka, "a stage on the way to art" (Heywood 75).

What connects the peculiar performances of these puzzling *saltimbanques*, therefore – be they hypothetically in a public square in Prague, under a *loggia* in Florence, or on a side-walk in Paris – is that a process of self-consumption through fasting or burning is the premise for creation, and that in both Kafka's and Palazzeschi's works "the author [has become] incorporated into [the] text through burning away his physical materiality and giving birth to a text that has consumed him" (Heywood 77). Even from an historical perspective, after all, an art and an aesthetics that intended to sever itself radically from the past necessarily had to purge itself of the "heavy" material residues of that past in order then to strive for a "higher" and "purer" art. As Marinetti put it: "Siamo imprenditori di demolizioni, ma per ricostruire. Sgombriamo le macerie per poter andare più avanti."[46] Moreover, a cyclical process whereby the "old" has to disappear in order to leave space for the "new" also seems to require (as "Boccanera" attests) that some sort of sacrificial rite of passage must take place, literally and figuratively, on the body of the subject as well as (in different measures) on the body of the text.

45 "How everything can be said, how for everything, for the strangest fancies, there waits a great fire in which they perish and rise up again ... *Only in this way* can writing be done ... with such a complete opening out of the body and the soul." (Anderson 251).

46 Qtd. in De Maria, *La nascita dell'avanguardia* (Venezia: Marsilio, 1986), 53. In this scenario, moreover, Marinetti's famous hostility to "pastasciutta" may acquire some extra layer of meaning (see Biasin 18).

This "act of disappearance" may present itself under many shapes and guises in Palazzeschi." Gestures such as stripping one's clothes off, or becoming transparent may thus be equally informed by a similar anorexic logic:

> La cosa più difficile alla quale possa un uomo arrivare è liberarsi di tutti i cenci, vestiti e vestitini, mantelli e mantellini, scarpe coccarde fiocchi, gale nappette e nastri, di cui gli altri lo avranno abbellito e coperto da quando era uovo nell'uovo dell'uovo. Curiosa faccenda, non è vero? Come si nasce vestiti. Parrebbe tutto il contrario. E che po' po' di lavoro, per quelli che ci riescono, potersi un pochino con grandissimo scandalo spogliare (In Tamburri 1990, 28).

It is a "hard task" ("che po' po' di lavoro") and a heroic act, especially in a society that, in Florence as well as in Prague, tended to privilege conformity and suffocate any real diversity. Whoever manages to succeed in such an effort, however, may achieve a status that is, again, a sort of "degree zero" from which it is possible to start over. One can even dream of stripping until nothing remains, achieving an almost incorporeal stage where it is impossible to say if someone exists or not – a point where, as in "A Hunger Artist," people may need to "[poke] into the straw with sticks" to check. It is only natural, then, to find transparency as a central theme in Palazzeschi's poem "Una casina di cristallo": "Io sogno una casina di cristallo/[...]/ma che sia tutta trasparente,/[...]/L'antico solitario nascosto/non nasconderà più niente/alla gente./Mi vedrete mangiare,/mi potrete vedere/quando sono a dormire,/sorprendere i miei sogni;/mi vedrete quando sono a fare i miei bisogni ...." Such a situation lets the crowd see and judge every aspect of the "live show" the poet puts on – a life whose arc significantly spans from a "full" (eating) to a "void" (evacuating). It is evident that such a concept also implies in itself the inherent possibility of invisibility and absence: seeing through something that is transparent and not seeing it at all are two faces of the same coin. The cuteness of the "casina di cristallo" and the jocose tone of the poem, moreover, cannot hide the fact that we are actually facing another kind of cage, another voyeuristic experience and, with all likelihood, another imminent instance of self-effacement for the sake of art, as the exclamation "Buon lavoro, poeta!" at the end of the poem seems to suggest.

Being progressively "lighter" up to a point where a status of "nothingness" is achieved is the paradoxical condition necessary both to survive aesthetically and to do one's job: to produce art. Once the artist has lost his *aureole*, s/he can only continue in a metaphorical stripping that becomes the pre-requisite to continue "heroically" to make art possible. S/he needs to be appropriately "reduced" to allow a much larger reality to take shape. The allusions to a body that constantly mutates and shifts in the direction of immateriality and transparency are clearly connected to a crisis of subjectivity, because they challenge the validity of rational perceptions and certainties traditionally associated with the idea of body and subject. As Brunella Eruli summarizes, such a crisis derives from the perceived gap between the world of history and the one of absolute values, and from the illusion that one could idealistically reconcile those two universes (Eruli 1970, 269–70).

The passages discussed above deal precisely with such an illusion, as they remind us that a fixed identity is impossible in the cultural and historical universe the two writers inhabit, and that disappearing may be just one among the many "masks" that one must wear. In this light, Palazzeschi's reference to clothing and stripping is an allegorical critique of the superimposed "vestments" of history (and, implicitly, of Western, bourgeois reason) and an implicit proclamation of the necessity of shedding them in order that the same subject may achieve a new creative status.

Both Kafka and Palazzeschi thus affirm some of the crucial "paradoxes of literary modernity" (Guglielmi 1991, 217): that the "degree-zero" (but also the origin, the center and the fulcrum) of subjectivity is a presence-absence, a full-void; that the state of absolute originality and the truest form of identity is the result of a performance, and thus an "act" or a kind of "true-fiction"; and, finally, that only an "absent" artist can eventually produce a "presence" even though this presence is constantly undermined by the ephemeral, by the contradictory, by the mutable.

Self-extinction and self-preservation are therefore only apparently antipodal in the logic I have delineated, as the former is a pre-requisite for the latter. The possibilities of a presence, of aesthetically surviving and, as a consequence, of being remembered, seem intimately linked to a situation which, either directly or indirectly, is close or conducive to an absence. That the idea of burning their books appears both in Kafka and in Palazzeschi seems to reinforce this argument, as a material object is eliminated so that its ideal and eternal part may continue to live.

To speculate on the reasons that "A Hunger Artist" is one of the few stories that Kafka asked his friend Max Brod to not burn after his death is likely to be more frustrating than illuminating. In the context of Kafka's desire to burn his works and his quest to "become literature" ("I ... am made of literature, I am nothing else, and cannot be anything else," qtd. in Heywood [76]), however, Palazzeschi's arsonist's claim that he will burn all exemplars of the first edition of his own poetry for a blessing, to make it truly incendiary, may offer some extra food for thought as well as reinforce the links between the two writers: "Là sopra il mio banco ove nacque,/il mio libro, come per benedizione/io brucio il primo esemplare,/e guardo avido quella fiamma, e godo, e mi ravvivo,/e sento salirmi il calore alla testa/come se bruciasse il mio cervello."

What is worth noting in these verses of "L'incendiario" is the implied identification of Palazzeschi's poetic self with the book he produced, as if he meant to "become literature" himself. The effects of burning his volume, in fact, are physically experienced by the poet's persona, the one who seems to actually be "on fire," as he feels the heat first in his head and then the burning of his brain ("la testa," "il cervello"), in anticipation of what shall soon be the fundamental quality of the "man of smoke."

Another, less popular poem by Palazzeschi, "L'ospite" (1910, in *L'Incendiario*), also connects the notion of self-preservation to the imagery of starving and burning. The poem asks the "Esseri umani alla terra aggrappati" what to do if the person they hate becomes a "guest" in their house, "come dentro una gabbia/dalla quale non potrà scappare," and the answer should now be eerily familiar: "Dargli fuoco!/Farlo morir di fame!/Guardarlo ridendo poco a poco/languire, struggere, consumare." However, this solution is immediately set aside ("Ma che!") in order to privilege a situation that cultivates a "personified" hate which is said to "vive[re] di sé": "Il tuo odio t'allungherà la mano,/gli porgerà il boccone./E all'ospite, il suo spirito di conservazione,/glie lo farà mangiare./E l'ospite rimane./Scrive alla sua famiglia:/ 'Sto bene, sono felice'./E voi l'odiate."

Thus, the acts associated with maintaining this uncomfortable status quo – feeding the guest and, on his part, eating the food – these instinctive acts, which stem naturally out of everybody's "spirito di conservazione," are also acts that "feed the hate" and keep each party involved in this binary dynamic (the host and the guest) in an emotionally stalling stage. From this perspective – the irrational, "contronatura" gesture of burning or starving the guest until he consumes himself – despite its sadistic cruelty, is a step necessary to "stir things up," to create a different and less static situation. Only by means of this kind of drastic dystopian scenario, marked once again by a "physical" vanishing, can one create favorable conditions for a radical existential transformation. In such an *a-topos* not only would hate no longer rule (or, perhaps, would just become a "productive" factor), but one would be able to overcome the stifling emotional spiral generated by the instinct of self-preservation and, with a Nietzschian twist, manage to lighten whatever load keeps men heavily "aggrappati alla terra," and thus achieve a "higher" and less material goal.

Perelà's situation in the novel named for him is strongly reminiscent of the one described in this often overlooked poem. The "man of smoke" himself is no more than a very particular kind of "guest" (*Il codice* 28; "Il re à ordinato che Perelà sia ospitato con ogni onore ..."), and one whose continuous presence eventually breeds a growing hate: "Tutti lo odiavano, e bisognava farlo odiare anche dal popolo [...] fu odiato, così odiato come nessuno fu mai" (157–8). This hate inspires the now familiar reaction of the crowd, which has decided to try and judge him: ("il taglio della testa [...] lo risotterrerei [...] gli preparerei il rogo [...] lo spoglierei sulla pubblica piazza e lo frusterei [...] lo metterei in un gabbione e lo farei vedere per curiosità" [175–81]). Perelà's reflections following the sacrifices of Alloro and the Marchesa Oliva di Bellonda are even more revealing: "Ma allora essi ànno ragione di odiarmi, gli altri, se amarmi vuol dire soccombere, ànno ragione, obbediscono al loro istinto di conservazione" (167). In a typical, paradoxical reversal ("to love means to die, so to hate is the only way to live"), they reaffirm the connection between hate and self-preservation established in the poem "L'ospite."

I find even more interesting, however, those passages in the novel that contain specific references to eating and food. Consider, for example,

Perelà's comments on the war, which he sees "come un'enorme minestra grigia, scodellata con stridulo crocrolo sciulo frastuono, e rimasta lì ... immangiabile" (19); or the Contessa Cloe Pizzardini Ba's sexual allusions: "Io mangio di buonissimo appetito almeno quattro volte al giorno ed il resto ... mi capite? E come non potrei pensare di rimanere una intera giornata senza prender cibo, non potrei pensare di rimaner senza ... mi capite?" (51–2); or the description of a typical (and dysfunctional) royal lunch: "Al pranzo di corte il Re non mangia mai, preferisce intrattenere il gentiluomo di destra, e lo manda via colla fame in corpo, così lui à mangiato prima, l'altro, povero diavolo, va a mangiare dopo, e questo è il pranzo di corte" (91). All of these passages corroborate the presence of an alimentary motif in the novel.

Two other passages are even more significant for my discussion:

> [Perelà] abbassò ancora gli occhi giù sull'ammasso enorme scomposto di giallastro, rossastro verdastro che formava il panorama della città e si sentì in quel momento di disprezzarla come preso da una nausea naturale, quell'ammasso gli appariva una vomitatura del padre eterno, dopo un suo pranzo ... da padre eterno [...].
>
> Gli uomini [...] quando vogliono esprimere il loro disprezzo, quando vogliono gettare in faccia ad un essere odiato l'insulto più atroce, si servono di ciò che di più intimo custodisca il loro seno (159/195).

The anorexic logic I have adopted as a key to the texts re-emerges here in two potentially related variations. In the first "sacrilegious" excerpt, it is God's gluttony that initiates the creative process.[47] Here, to be precise, bulimia takes the place of anorexia as a metaphorical means of creation: "il panorama della città" is the result of a divine "binge-purge" cycle, as if to imply that even God, in order to "create" had to make himself "lighter." The "divine" regurgitation of food is a form of creation and may constitute a first allusion to the parallel process performed by the "god-like" writer, the one who is actually "throwing out" and "regurgitating" his words on the page.

The second passage further supports this interpretation, as it establishes a subtle but tangible connection between men's capacity to express

47 On "God's belly" see (Camporesi 73).

scorn and hate and the actual, physical "expressing" of some sort of organic matter.[48] What these lines seem to imply – especially if one looks at the expression of hate as a paroxistic allusion to human expression in general – is that the acts of communication and of producing meaning, on the one hand, and the acts of "emitting/expelling" the most intimate "material" of one's heart/body ("seno"), on the other, may actually constitute the same thing for the narrator of *Il codice di Perelà*.

The implications of these poignant excerpts in Palazzeschi's text are interesting. On one level – as both passages reaffirm – creation, be it divine or human, requires more or less radical "lightening" or "emptying" of the self. Metaphorically, for Palazzeschi, to communicate or, more specifically, to write, means to release the inner part of himself and, since one becomes "lighter" in the process, this act can be interpreted as a sort of "digestive" gesture. Remember once again *Perelà*'s dedication: not only has the author received "frutti" and "verdure" from his scornful audience but, at the conclusion of the book, he has processed that same organic material and expelled it, throwing it back in order to "cover" them under a metamorphosized, improved, and "delicious" shape. It does not come as a surprise to find out that Perelà himself (who, with his enigmatic Code, can be seen as the ultimate meaning-producer and artist-figure) is the peculiar by-product of digestion. Before his "scendere alla luce" and his expulsion from the "utero nero," he describes himself in terms that can be best defined as "pre-fecal": "mi svoltolavo da ogni parte come un globo di una pietanza indigeribile in uno stomaco umano" (27).[49] "Normal" human beings, on

48 See the "sput[i] [...] lucenti, grigiastri, gialli, enormi che si vanno a squagliare in terra come chiari d'uovo" (195) when Perelà walks by, and, even "worse," the whole anticipatory episode of Iba: "da tutte le finestre di tutte le case di tutta la città piovve su lui nelle più svariate maniere la stessa cosa! [...] Forse egli è felice, affonda la sua oscena pancia nell'immondezza colla quale fu incoronato" (116).

49 This interpretation, in turn, makes one wonder about the potentially vulgar nature of the "fumo" Perelà is made of. The fact that Camporesi, writing about Dante's "Inferno" XXI, speaks of "anal wind as a function of the genetic proliferation of new lives [which] reflected with grotesque efficiency old beliefs concerning the anal birth of souls" (Camporesi 18), may positively add to my observation.

the other hand, all those gravity-bound characters who constitute the vociferous crowd in the novel (with the notable exceptions of Alloro and the Marchesa Oliva di Bellonda), those same ones who "la terra gli attrae" (33),[50] are said to be and to remain indigestible by the ground: "La terra gli vomiterebbe ben volentieri tutti all'infinito! Povera terra, sono la sua pietanza più indigesta, gli à tutti fra la gola e lo stomaco, nessuno le è passato ancora nell'intestino" (33–4).

In direct contrast to Perelà's fasting habits ("non mangia"), the "earthbound" characters possess an insatiable hunger ("la fame di quei feroci animali è insaziabile" [166]), which further explains their hateful behavior toward the protagonist and, arguably, their overall worldly weight. The most important consequence of not being "vomited" by the ground and, thus, of not "descending to the light," is that they literally and metaphorically "remain in the dark," that is, they are unable to see and recognize Perelà's Code and his philosophy of "leggerezza." Their inability to be "digested" or expelled, therefore, is directly linked to the impossibility to truly understand and communicate, two acts which, ultimately, are also the necessary ingredients for any artistic creation.[51]

It is only natural, then, that within the same symbolic frame of reference, Perelà himself is described, again in culinary terms, as "delizioso" and "squisito": ("È l'uomo più squisito ch'io m'abbia mai conosciuto [...] quel delizioso signore" [47]). As such, Perelà recalls the "delicious" "opere d'arte" with which the writer, in his dedicatory note, promises to "cover" the public.

But there seem to be inherent dangers in the ambitious creative process I just described. That same writer who has been able, with alchemical skill, to transform "frutti e verdure" into "opere d'arte," who has managed to process such a food and then expel it in the form of a text, is also telling his readers about the risks associated with such an act, as the final, vanishing

50 The inter-textual reference to those humans "aggrappati alla terra" in the poem "L'ospite" is evident.

51 In an analogous way "fasting [is] a condition hardly to be understood by well fed people" (Kafka 1974, 272). A variation of this problematic, as it relates to the capacity of remembering, resurfaces also in F.T. Marinetti's novel, *The Untamables* (1922).

gesture of his protagonist may suggest. That is, there is the risk of being consumed by the creative effort and the danger that art may actually fail to change society positively. Yet it remains the case that such a self-effacing gesture is the only means of achieving artistic success and exorcising potential failures. Erasing oneself through burning or fasting is, in fact, necessary in order to "burn off [...] the failed attempts to reach the originary purity not possible in daily life" (Heywood 81).[52]

To suggest that Palazzeschi's *Perelà* and some of his earliest poems are partly informed by a variation of the "anorexic logic" that is also operational in several of Kafka's texts and that the "omino di fumo" is also an "omino di fame," is as close as it is possible to approach my goal. Even though Palazzeschi did not leave a series of letters and diary entries in which he analyzed himself and his obsessions in painful detail and he did not explicitly write – as Kafka did in his famous diary entry of January 3, 1912, when he wrote that in order to create he had to "diet in all directions," his literature of "leggerezza" speaks for itself.

Indeed, the connections between Kafka and Palazzeschi may perhaps contribute to the delineation of a potential *trait d'union* between European modernism and the Florentine avant-garde. For both Kafka and Palazzeschi, any meaningful act of creation and communication is a deeply existential (and "physiologically necessary") activity, in the (limited) sense that, like burning, eating, and/or fasting, it consumes oneself, and thus deals simultaneously with the possibility of life and death. It is a "trial" that may end up annihilating its creator. Especially for these two writers, it is a passionate act that involves self-sacrifice and that stems equally from love and hate. Writing, like eating, is the obvious path to survival and self-preservation, but the inevitable paradox, in Prague as well in Florence, is that, at a tropological level, in order to accomplish the former one has either to eliminate the latter or re-invent its meaning (i.e., "mangiare il fuoco").

52 "Non è egli [Perelà] la sublimazione del corpo e dello spirito umano? Non viene egli quasi a darci prova di altri destini, di altra vita, vita e destini, nei quali gli umani egoismi, gli umani traffici non ànno più la loro parola?" (*Perelà* 83).

If in the past the ingestion of food could translate into a Hegelian desire for "assoluto dominio e possesso" and "impossessamento e trasformazione del mondo" (Biasin 25–6), in the European modernist context represented by Kafka and Palazzeschi, such an objective is not only questioned in its theoretical principles, but its potential and dubious achievement may be reached – if it can be reached at all – only through the opposite, "negative" gestures of fasting and burning. Perhaps, the best thing about this "starving" scenario, one where ultimately "the creation of the work of art demands the deconstruction of the body" (Ellmann 69) is that, in the meantime, fiction and poetry writing can hardly get any more appetizing.

2.3 Starving to Get Full: Aldo Palazzeschi's *La piramide*

> Si è detto che il mondo è come un'arancia. Giustissimo. E tu ti contenti di mangiartene uno spicchio? Cucù! A me fa gola tutta!
>
> — A. PALAZZESCHI, *Equilibrio*

> Books are not made like children but like pyramids.
>
> — FLAUBERT TO FEYDEAU, 1858

Aldo Palazzeschi's early works contain a variety of evocative architectural images. Decadent palaces, churches, ancient temples, old monuments, sanctuaries, and fairy-tale dwellings provide the setting for many of the poems in *I cavalli bianchi* (1905) ("Il cancello," "Il tempio pagano," "Il manto," "Il castello dei fantocci," "La casa di Mara"); in *Lanterna* (1907) ("Torre burla," "Tempio serrato," "Palazzo Mirena"); and in *Poemi* (1909) ("La porta," "L'arcario," "Il tempio"). If, on the one hand, these images suggest the writer's fascination with a *crepuscolare* repertoire of haunting buildings, on the other, these edifices seem also to stand as ephemeral emblems of stability within a poetic landscape that progressively undermines any illusion (or actual wish) of solidity and permanence, be they architectural or existential.[53]

Moving from poetry to prose, it is not surprising that after the burned-down Villa Bemualda at the end of the first part of *: riflessi* and the emptied prison cell at the conclusion of *Il codice di Perelà*, an "invisible" pyramid, present as it is only as an enigmatic linguistic reference in the title, constitutes the symbolic foundation of Palazzeschi's third "extraordinary novel": *La piramide. Scherzo di cattivo genere e fuor di luogo*, written in 1912–14, but not published until 1926.[54]

The novel is divided in three chapters entitled respectively "A tre," "A due," and "A solo." In the first one, the self-reflexive protagonist and first-person narrator finds himself at the Florentine *belvedere* of Piazzale Michelangelo next to two other men, who are soon revealed as an optimist and a pessimist. The former extols and celebrates the wonders of nature and the whole universe, while the latter embodies an antithetical, deprecatory position of *contemptus mundi*. The protagonist's attempts to establish some sort of closeness, friendship, and communication with both of them fail and, in the end, he finds himself alone. As he says "Io, non potendo essere andato né con quello né con quell'altro, rimasi [...] lì, lì così, sì." (84).[55]

53 When discussing Palazzeschi's poetic beginnings, Mimmo Cangiano aptly observes that "solo un'arte come gioco, un'arte non proiettata al fondamento di valori stabili, potrà non tradire la fluidità della vita." Mimmo Cangiano, *L'Uno e il molteplice nel giovane Palazzeschi (1905–1915)* (Firenze: Società Editrice Fiorentina, 2011), 43.

54 The pyramid well symbolizes the existential extremes (i.e., heaviness and lightness, surface and depth) within which many of Palazzeschi's fictional characters traditionally move. Its chthonic, massive stone base solidly grounded on the surface of the earth contrasts with its increasingly thin apex, which vanishes vertiginously into the air. As its exterior surface covers the mysteries of its interior depths, one remembers that this is also a tombstone, an exotic and peculiar emblem of death and, at the same time, a portal that allows those buried inside to eventually begin a new, immortal life in the beyond. Life and death, beginning and end are immediately evoked by the architectural structure itself.

55 Guglielmi notes that "le due figure sono portatrici di parole fondatrici, di visioni globali del mondo." Guido Guglielmi, *L'udienza del poeta. Saggi su Palazzeschi e il futurismo* (Torino: Einaudi, 1979), 80. Terrile adds that they "vengono ad assumere

In "A due" the protagonist then confronts and, once again, fails to positively resolve an intricate system of *alter ego* that presents itself as a "treasure" represented by six imaginary friends (Cecco, Pietro, Agesilao, Gaspero, Giacomino and Tiberio) with whom he has a series of surreal encounters and ultimately unsatisfactory conversations. Guglielmi speaks of these additional friends as the anthropomorphized projections of his inner drives (Guglielmi 1979, 82), Terrile describes them as "rappresentazioni soggettive" (114) and Cangiano refers to them as the "simulacri [...] proliferazione di maschere identitarie" (182).

Finally, in "A solo," we encounter the protagonist, now completely alone, who punctiliously organizes travels to various destinations only to renounce actually taking the trips, thus ending the novel and his "opera di demolizione del mondo" (Guglielmi 1979, 84) with a paradoxical and ironical "elogio dell'attesa," and the apparent value assigned even to a "negative" desire (Cangiano 191–3).

One may further summarize the existing scholarship on *La piramide* by noting that the several initial and generally not very generous critical responses that used terms such as "involuzione," "divagazione," "rallentamento" and "sfocamento" to describe it in comparison with Palazzeschi's previous novels, were substituted, as time passed, by more accepting and nuanced interpretations (such as those by the scholars mentioned above) which revealed not only the relevance (and the failures) of its linguistic and stylistic aspects, but also the complexity, interconnectedness, and layered allegorical structure of the text.[56]

[...] lo statuto di discorsi letterari." Cristina Terrile, "*La piramide* di Aldo Palazzeschi: Un'allegoria del discorso letterario," in *The Italianist*, 20 (2000), 106.

56 I am referring in particular to the commentaries by L. Orsini, "La parabola di Palazzeschi," in *La cantina di Auerbach. Sette Saggi* (Napoli: Edizioni Scientifiche Italiane, 1971), 9–95 and Marco Forti "Romanzi straordinari", *Palazzeschi oggi.* Atti del convegno Firenze 6–8 Novembre 1976, a cura di Lanfranco Caretti (Milano: Il Saggiatore, 1978), 111–32. One notable exception was Eugenio Montale who wrote that *La piramide* "è un libro che se fosse stato scritto da un esistenzialista francese avrebbe fatto il giro del mondo" (in Cangiano 193). For a more complete overview of various disappointed critical responses to *La piramide*, see the introductory section and the relative notes of Terrile's article.

Guido Guglielmi, for example, in his *L'udienza del poeta* (1979), was among the first to focus on the text in its whole rather than on the discontinuities between its three chapters and, therefore, to provide a more comprehensive reading of it. After noting that the novel is "la parodia di un viaggio di iniziazione" and that "come metafora visiva la piramide è la figura di un'ascensione progressiva, da una base a un vertice, da un fondo a una cima, dal più al meno, il simbolo di un processo di illuminazione che coincide con un processo di riduzione, assottigliamento e progressivo annichilimento del mondo" (78), he suggests that such a travel is ultimately one of the self in search of the "invenzione di una nuova idea e di una nuova presenza della poesia" (83).

Cristina Terrile, in turn, expands on Guglielmi's intuition by interpreting the novel in the light of a general desecration of the literary discourse and, at the same time, of the vital need to express oneself through a new kind of language and a different sort of representation. Such a desecration, she also observes, is primarily achieved through the widespread use of coprological metaphors. In the first section, the literary discourse "appare come strumento di una verità che può essere condivisa," while the second and the third sections mark the shift to "un uso soggettivo della letteratura e delle rappresentazioni che essa alimenta" (102). Once traditional literary discourse has been recognized as impossible, the poetic self at the end of his journey refuses to consider representations as the exact image of the reality in front of him and, instead, accepts them for what they are, that is, not "come vie d'accesso ad un'altra realtà, ma come rappresentazioni singolari, perfette e eternamente rinnovabili della soggettività" (114).

In general, the more insightful interpretations of the novel agree on the meta-literary quality of the text[57] and on the fact that its ultimate message includes the following, interconnected themes: the affirmation of the freedom of creative imagination (Guglielmi 1979, 84–5); the search for a personal literary language and voice "capace di esprimere la soggettività" (Terrile 115), and the escape from the "petrifying" quality of words and

57 See Luigi Fontanella, *Il surrealismo italiano* (Roma: Bulzoni, 1983), 102.

thoughts towards "le infinite possibilità che la fantasia prospetta e che la vita contiene" (Cangiano 182).[58]

My own objective in these pages is to build on these critical premises in order to make some additional reflections on the presence of the text's coprological and alimentary references. I wish to argue that these references do not simply desecrate and lower the register of the discourse, but also, as they relate once more to both the processes of renunciation and of "negative" fulfillment of the protagonist, may help to corroborate from an additional angle the metaliterary quality of *La piramide*.

In particular, especially in the light of the already noted "process of reduction" in the text and of the progressive pattern of renunciation that the protagonist embarks upon, I would like to argue that an examination of the representation of the alimentary imagery in *La piramide* reveals that this novel too is characterized by allegorical connections between gestures of fasting and starvation and questions of aesthetic creativity, which are reminiscent of those in Palazzeschi's earlier texts.

Because, from its first pages, *La piramide* clearly addresses the problematic connection between words and things, abounds with gastronomical metaphors of incorporation and ejection, and displays successive gestures of sacrifice and renunciation that culminate with an enigmatic laughter, the previously introduced notion of "literary anorexia," and the realization that, as Mark Anderson noted, "one could not talk about a 'language crisis' in European literature at the turn of the century without also addressing the problem of food rejection [...] [and] figures of anorexia" (Anderson

58 This need to escape from the "prison-like" constrictions and conventions of standard language and reach a point where the body as well as the word is "emptied out" of its usual meanings and, therefore, open to be filled with all the remaining potentialities of communication strongly recalls some of the positions of Georges Bataille. For example, when Bataille writes that he is "betrayed by language," that is, by its "strettoie [...] che [...] ostacola[no] le erranze di un pensiero funambolico" (qtd. in Carlo Pasi, *Georges Bataille, La ferita dell'eccesso* [Torino: Bollati-Boringhieri 2002], 231), he seems to be sharing with Palazzeschi the needs to resist the homogenizing, totalizing, and limiting characteristics of language and to achieve the state of existential "divertimento" (also in the sense of "detour" from the norm) typical of his *saltimbanco*.

29), would again seem useful to demonstrate that here, too, we are facing an additional instance of self-consumption that opens up the possibilities of aesthetic creativity.[59]

At the conclusion of *La piramide*, the protagonist achieves an ideal status of receptivity, one where his chosen condition of "un-fullfilment" becomes the paradoxical premise to be then filled (or enlightened) by all the possibilities of creative imagination ("a pigliare [...] un po' di sole").[60]

As noted, the relationship between things and words and, in turn, the one between words, food, and the body is addressed at the very beginning of the novel. The text begins by comparing vowels to cherries that one may alternatively ingest or spit out ("queste vocali [...] l'una tirando l'altra come le ciliege [...] aspirare una vocale! che è mai? gettarla via dalla bocca!" [7–8]), then allusively describes the path that such vowels may take to produce two different sorts of discourses:

> [...] affinchè una volta presa la spinta dal basso la vocale [...] incominciata a venire su, non dovesse rimanerti rassegata nella gola [...] [or] accade anche questo, per isfogare da tutt'altra via, recondita, lunghissima ed oscura, e che non ha nome, o se l'ha non si dice. (9)

The novel ultimately returns to a hilarious "curiosa metafora coprologica della creazione letteraria" to reveal that both creating and defecating are part of the same process, that is, a "ponderosa fatica di scioglimento" (Terrile 101).

59 Interestingly, Cirlot noted the symbolic association of the pyramid with the notion of creation: "There is an apparent contradiction in the symbolism of the pyramid [...] it is symbolic of the earth in its maternal aspect [...] [it] express[es] the twofold idea of death and immortality [...] [it is] a synthesis of different forms, each with its own significance. The base is square and represents the earth. The apex is the starting-point and the finishing-point of all things – the mystic 'Centre' [...] Joining the apex to the base are the triangular-shaped faces of the pyramid, symbolizing fire, divine revelation and the threefold principle of creation. In consequence, the pyramid is seen as a symbol expressing the whole of the work of creation in its three essential aspects." J.E. Cirlot, *A Dictionary of Symbols* (New York: Dorset Press, 1971), 268.

60 Terrile individuates the unity of the text in the protagonist's "disponibilità a divenire di un essere essenzialmente svuotato." (Terrile 114).

> Ci sono uomini che per fare una certa cosa, dopo vi dirò quale, incominciano dall'indossare un'apposita uniforme, entrano con misteriosa cautela in locale risplendente di ceramiche e di metalli, e luce elettrica, adornato di recipienti svariatissimi e curiosi, apparecchi di caricamento e scarico, tutti gl' istrumenti della chirurgia; e una volta che vi si sien rinchiusi van nel mezzo a sedervisi come il papa in soglio; e ti fanno scatenare, dopo un tragico silenzio pieno di raccoglimento e di mistero, nel quale tu pensi, o ti domandi, quello che mai saranno andati a fare, dei rumori così orrendi, che pare che abbiano dato via al demonio: e crosci di cannelle, pannilini, saponi d'Inghilterra, acque di Colonia, cipria, unguenti, estratti di Parigi, prima che ne sien fuori ce ne vuole. E ve ne sono, che dove si ritrovano, poco si fanno in là, buttano giù i calzoni e: *pataplum:* "Ah!", od una pigiatina tutt'al più: "*hum*", due umilissime foglioline che possano per caso lì trovarsi, che non sieno d'ortica ben s'intende, o un sasso, non a punta, si capisce, con maggiore abilità ... e meno spesa, ritiran su i calzoni e dopo un gran sospiro di sollievo se ne vanno guardando il cielo; e fan la stessa cosa. (13)

This is the specific passage in which, as Terrile observes,

> il discorso letterario viene assimilato ad un comunissimo processo fisiologico, che soddisfa principalmente colui che lo espleta [...] come non è possibile vivere senza defecare, non è possibile [...] vivere senza produrre discorso e rappresentazione." (102)

These references to the physiologically unavoidable acts of writing and, alternatively, those of ingesting or expelling foodstuff continue to recur as the novel proceeds. Thus, in "A tre," the words the first optimist "amico" uses to poetically praise the moon become "bazzeccole [che] se le baloccolava colla lingua in bocca come caramelle" (22); the moon itself is imagined as a "polpetta" (23) and, a few pages later is described together with all the other natural wonders traditionally chanted by poets (the sun, the plants) as "robe [...] più vecchie della broda dei fagiuoli" (37). Additionally, when the "pessimist" appears, one finds that "guardando masticava [...] biascicava soltanto delle parole [...] Digrignava i denti stritolandovi dentro le parole" (38–9). The pessimist's whole discourse is so rich with negative alimentary allusions ("la felicità ci sarà ma se la papperanno tutta loro," "E finirai in cucina, tra le ganasce rabbiose della sguattera,") that the protagonist/narrator, in turn, admits that he could feel "rivenire alla gola tutti gli stufatini e scaloppini al Madera, bistecche, rostbeef, filetti alla parigina e alla salvietta, involtini e saltimbocca" or, explicating the metaphor, that he cannot really stand the over-elaborated and hard-to-digest literary discourses belonging to a past literary tradition.

In "A due," the chapter in which the narrator confronts his "proiezioni" (Guglielmi 1979, 82) or "emanazioni fantasmagoriche" (Terrile 109), the food metaphors substantially decrease. However the alimentary motif is also present here in the sense that a few of such projections emerge out of (and then disappear again into) the protagonist's body as if they were, alternatively, excremental or edible matter. The following, for example, is what happens when Cecco appears: "un brivido scosse tutto il mio corpo [...] e parvemi che [...] un altro corpo uscisse dal mio corpo stesso" (99), and then vanishes: "Con quel brivido istesso che me l'avea fatto dar fuori, ora io me l'era ringoiato" (102). Likewise, when Agesilao appears, the process is once again described as associated with the act of evacuating oneself: "come si fosse sradicato dalle mie più riposte viscere [...] un coso lungo, intirizzito, secco" (114). After Giacomino's appearance, instead, the protagonist feels like eating: "Non vi sentite dopo, voglia di mangiar qualche cosa? Di buttar giù? Io sì, ve lo confesso." (128).

What these references all seem to suggest, beyond their stylistic flamboyance and (homo)sexual allusions, is that in order to proceed towards his final destination and find what he is looking for, the protagonist needs to go through something resembling a digestive process. He must eliminate whatever is superfluous and old in the world and within himself and, at the same time, re-elaborate the useful, constructive parts.

The third section of the novel, "A solo," provides best verification of the applicability of the previously introduced notion of "modernist anorexia" to *La piramide*. In these pages the existential cost the protagonist has to pay to reach the end of his journey and find a new voice emerges more clearly. This is the chapter in which the protagonist, now alone, elaborates at length on the idea that the most perfect and satisfying journey is the one never taken, the one that remains at the organizational and planning stage, at the level of a perennially unfulfilled desire. As a consequence, it is more rewarding for him to prepare for and dream of journeys to various locations (Venice, Rome, Naples, and Egypt) than to take the actual trips. Reality for him is always inferior to the imagined.

In line with Palazzeschi's "topsy-turvy" poetics, the downward, southern direction of the planned trips (from north to south, Venice to Egypt) clashes with the general direction needed to reach any sort of upward and, therefore, instinctively "northern" summit. That is, as one expects the protagonist to move upwards, towards the elusive apex of the pyramid, the text counterintuitively presents us with an imagined series of travelling plans that – precisely in order to reach that "high" goal – need to go in a lower direction or southward. This situation seems to reiterate through a different metaphor what was expressed in the previously quoted coprological passage. In other words, just as one does not necessarily need an upscale marble bath "risplendente di ceramiche e metalli" to fulfill one's physiological necessities (i.e., the need to artistically create) and may achieve the same objective wherever one finds him/herself ("dove si ritrovano [...] fan la stessa cosa"), true literature does not always require an "upward" search for a particularly elaborate and elevated language but could also be made of humbler components, such as everyday words.

This said, my attention is especially drawn to the path of physical sacrifice and renunciation that the protagonist embarks on in order to go to Egypt (with the excuse of saving money for the trip), since the bulk of his actions has to do with the drastic reduction of his food intake: "dovrò sottopormi durante mesi e mesi alla tirannide di economie quotidiane ristrettissime [...] Abolita la frutta, i dolci, il formaggio nella minestra, ogni possible ghiottoneria bandita, bandito ogni svago, ogni lusso; dovrò forse bere acqua sei giorni della settimana." (150) and, a few pages later: "sto facendo economie atroci, non più frutta nè dolci nè un fiore, più nulla, invece di un Venerdì la mia settimana ne ha tre, Quattro, ne avrà sette se occorreranno [...] sono proprio queste crescenti e reali sofferenze fisiche e morali che mi procurano sempre più grande il premio del mio bellissimo viaggio." (158)

Abstinence from food, an increase in the number of fasting days ("Venerdì") during the week, and physical starvation are the gestures that paradoxically allow the protagonist to obtain his "prize" and move towards his ultimate, imaginary destination: an "'ideale' da non raggiungere mai" (Cangiano 191) which also coincides with the more tangible conclusion of the text. The "philosophy of incompleteness" that informs the last chapter of *La piramide* thus associates the "emptying," "starving" gestures of the protagonist with the progressive fulfillment and strengthening of his imaginative powers (coinciding with the material end of the novel).

As a matter of fact, before the protagonist can weave his own clothes ("tess[ere] le mie tele") and work on his own project ("architettare io sull'architettura delle Piramidi"), he informs us that his own planned journeys are based on the metaphoric "ingestion"[61] of previous oral, written, and visual sources that he needs to thoroughly research and explore:

> Ma certo, certo, io ho una cognizione abbastanza approssimativa dei paesi che sono per visitare, per avere udito narrazioni, letto storie e leggende, veduto illustrazioni, e sempre la coltivo ed è appunto su questo lavoro ch'io tesso le mie tele, e baso la mia felicità. Voi non penserete ch'io mi accontenti di vedere la Sfinge o le Piramidi tali e quali si vedon sui panetti della cioccolata di Talmone, ma giungo a ricercare tutti i poeti che le hanno cantate o descritte, i pittori che le hanno dipinte, e tutto serve come punto di partenza per architettare io sull'architettura delle Piramidi. Le Piramidi sono fatte a punta e non come i cocomeri, per voi e per me, e per tutti, ma solo che la punta delle mie Piramidi non è la vostra nè precisamente quella dell'Egitto, nè, tampoco, quella della cioccolata di Talmone, ma è una punta solamente mia. (160–1)

In a second moment, however, we also learn that such a tradition ("narrazioni [...] storie e leggende [...] illustrazioni") needs to be reconsidered, presumably after his having assimilated and digested it ("cose viste e riviste le migliaia di volte, più vecchie della barba di Noè, e di cui sono ricolme milioni di scartofie rognose," [181]), and then literally expelled it ("scarica[ta] all'aria libera" [181]), in order to be able to achieve that "punta solamente mia." Just as the ingestion and elimination of food is physiologically necessary for one's existence and well-being, the active incorporation of past "narrazioni [...] storie [...] leggende" is necessary for finding one's own voice and, ideally, one's happiness. At the same time, however, the text suggests that such a destination cannot be achieved without a metaphorical "emptying" of oneself which ultimately, echoing a similar situation in *Il codice di Perelà*, prevents one from remaining "stuck" in an unresolved and nonconducive middle ground or, as Palazzeschi explicitly suggested

61 "Ingestion" would seem an appropriate term, given the culinary references in the passage to pyramids on chocolate bars and "pyramids-as-not-being-watermelons."

in his 1915 manifesto *Equilibrio*, "a mezza via [...] vittima di tutto il sego che imbottisce lo spirito umano." Accordingly, the need for various kinds of "sciogliment[i]" (10) is repeatedly expressed throughout the novel:

> mi strappavo i panni di dosso non sapendo se mi dovevo tirar giù, o se mi dovevo tirar su' (29); Si può seguitare a vivere così, sempre nel relativo, senza la più vaga speranza di poterne uscire (78); Gustare cose, nella vita, a metà, mi sembra contentarsi troppo presto, rimanere a mezza via vuol dire in fondo non possedere la capacità di andare avanti (104); rimaneva lì piantato, quasi di sale, senza la forza necessaria per andar nè in su nè in giù. (115)

Creative originality and a full poetic identity cannot be obtained by continuing to regurgitate stale poetic words contained in heaps of old papers. Instead an "anorexic" process of abstinence and renunciation that entails both the elimination of the old and the generation of the new is required.

In this light, it is only a false paradox that the embodied exemplification of such a philosophy is said to be *Sverze* "il pezzente." The character who has made a way of life out of renunciation ("Gli hanno offerto posto e lavoro, ha rifiutato; cibo e ricovero, ha rifiutato") and who, significantly, is not afraid of being hungry ("la fame non gli fa paura") is also the first figure able to find, stop, and hang on to the "punto" of his "happiness" ("C'era il punto di essa, lui l'ha trovato, ci si è saputo fermare e lo tiene" [176–7]).

Not only, therefore, do happiness and satisfaction stem (with a Leopardian echo) from the deferred fulfillment of a desire but, in turn, this elusive fulfillment derives only from a series of renunciations. Desire, in other words, keeps the world full, and renunciation, in the end, prevents that same world from becoming empty. As the protagonist remarks:

> Colle venticinque lire [...] volli comprare quella famosa caraffa di Boemia [...] ma una voce in fondo mi diceva: "non la comprare".
>
> [...]
>
> Se avessi poi tutte le volte il denaro in tasca per comprare seduta stante una cosa che sul momento mi piace, e facessi l'errore madornale di comperarla, il mondo si vuoterebbe d'un tratto per me. (171–2)

Even the protagonist's economic behavior indicates an attitude of sacrifice and renunciation. And since the term "economy," besides meaning "management of money, or of the finances," also refers to "the rules that control a person's mode of living; regimen, diet" (*OED*), one cannot help reflecting on this apparent link between the economic and the gastronomic.

If it is true that refusing to eat and refusing to buy are part of a similar rhetoric of renunciation, it may also be true that these two gestures share the common goal of undermining and substituting an old, functional model of full capitalist, consumerist consumption based on an indiscriminate ("bulimic") accumulation and ingestion with a new "counter-diet" a dysfunctional physiology based on abstinence and fasting. That is, these acts simultaneously establish an alternative economy of speech, a new kind of discourse, and an alternative model of modern subjectivity which "non cerca di serrarsi in un'unità, perché vive questa come una limitazione delle proprie capacità" (Cangiano 187). Rather than risk being stuck in an unsatisfying and false sense of fullness, the protagonist finds more rewarding and liberating to "avoid the purchase" or, in parallel, to "starve" since both acts offer an (allegorical) escape from a traditional, paralytic creativity and, at the same time, irreverently undermine one of the strongholds of the bourgeois way of life.[62]

In this scenario, the final burst of laughter of the protagonist as he reaches the (imaginary) apex of the pyramid stands out not only as a potential reminder of the ever-present motif of a clownlike lightness, but also as the latest, poignant symptom of "bodily discharge" that sanctions the transition from a previous stage of heavy mental and physical constipation to an imaginary, liberating and "relieving" spatial-temporal coordinate, where a different kind of discourse will be able to loosely flow and a new kind of art will be able to flourish.

At the conclusion of the novel, after a strenuous path marked by renunciation and sacrifice, the protagonist finally appears to have reached this

62 Once again, it is tempting to establish a connection between the peculiar "philosophy of renunciation" (and, therefore, of implicit "loss") expressed at the conclusion of the novel and Bataille's elaborations on the notion of *dèpense* which he intended as necessary to communication and a "distruzione senza profitti, che esalta la piena libertà dell'istante [...] dissipazione improduttiva [...] dinamica dello spreco [opposed to the] avidità possessiva della borghesia" (qtd. in Pasi, 210–22).

sought-after destination. He finds himself at the top of the pyramid, in that *Mittelpunkt* between earth and sky where, as Guglielmi observed, one simultaneously faces "the franamento comico del senso del mondo" (Guglielmi 1979, 87) and the realization that the journey may have actually just begun.[63]

As Palazzeschi suggested both in *Varietà* and *Equilibrio*, two of the earlier manifestoes that anticipate several of the arguments later hidden in the labyrinthine prose of *La piramide*, in order to attune oneself to the fundamental principles of (artistic) creation – which is rooted in difference rather than homogeneity and in plurality rather than singularity – one should conceive the "punto" where one is born (both biologically and aesthetically) at the center of the circle of life as an end as well as an origin, as a place of arrival as well as a point of departure.

Although the geometry is obviously different, it does not seem too far-fetched to suggest that the characteristics just delineated may belong also to the comparable *Zentrum* at the top of the pyramid. More specifically, Palazzeschi in *Equilibrio* writes of a situation of only apparent balance, of a "punto di attrazione" from which to depart and return repeatedly in order to reach, experience, and, "taste" in all their variegated depth, the range (or, if I am allowed the pun, "the o-range") of possibilities that, otherwise, would lie only "superficially" on life's "circumference." As he writes towards the end in *Equilibrio*: "noi non siamo rimasti al centro per cullare la nostra pigrizia. [...] Noi abbiamo rifiutato tutti i colori per avere la gioia di poterli assumere tutti quando ci piaccia [...] Si è detto che il mondo è come un'arancia. Giustissimo. E tu ti contenti di mangiartene uno spicchio? Cucù! A me fa gola tutta!" (30–1). And in *Varietà* he adds: "Quando una cosa è bella e fatta bene e vi piace prima cosa da fare sarebbe

63 I am borrowing the term *Mittelpunkt* from Aldo Gargani: "E questo *Zentrum*, questo *Mittelpunkt* è il luogo ideale in cui l'uomo è nuovamente capace di fare esperienza dell'etica e della conoscenza, finalmente sottratte al movimento, alla concatenazione delle cause e degli effetti, alla meccanizzazione logica degli stili intellettuali [...] una conoscenza [...] un'illuminazione [...] che è sottratta, per così dire, alle intemperie dello spazio e del tempo." Aldo Gargani, *Il coraggio di essere* (Bari: Laterza, 1992), 8–9.

di fuggirla per farne una differente se aveste davvero in corpo lo spirito della creazione" (6).

One of the solutions to the enigma of *La piramide* lies somewhere between these two interrelated affirmations. The former, in a quintessentially "anorexic" logic of renunciation, presents a paradoxical premise to a future existential situation of joy and satiety. The latter premise asserts that true creation despises uniformity and always prefers a loose "under construction" and unreachable project to a fully completed one.

The remark that ends the novel ("E per questo irto cammino, giunto io sono alla sommità della piramide: solo quassù. Chi sa cosa credete ci sia venuto a fare. Ah! Ah! Nulla, a pigliare … un po' di sole.") thus seems to indicate, again echoing Kafka, that solitude is a crucial condition to being able to create and to remain arbiter of the necessary receptiveness to do so and, echoing Bataille, that the protagonist is now emancipated from the "architectural prison[s]" of both a univocal identity and, correspondingly, a definitive, "calcified" language. Once at the summit, he is unburdened, "relieved" and ready to "have fun" by gorging on the infinite (and "labyrinthine") possibilities that his imagination and language, in all its registers and manifestations, shall offer him.[64]

2.4 On Voraciousness: Aldo Palazzeschi's *Interrogatorio della Contessa Maria*

More than two decades have passed since the discovery and the posthumous publication of *Interrogatorio della Contessa Maria* (from now on, *Interrogatorio*), the novel that Aldo Palazzeschi wrote around 1925–6 and

64 Bataille's negative view of architecture and the opposition between the pyramid as an image of "architectural prison" vs. the labyrinth as a Protean, atemporal, open structure is discussed in Dennis Hollier's *Against Architecture: The Writings of Georges Bataille* (Boston: MIT Press, 1992).

then decided to leave unpublished.[65] But only a few critical studies address this unusual text. Most of these do not venture much beyond the description of its irreverent, erotic content, or the individuation of motifs associated with the poetics of "lightness" that Palazzeschi displayed in his previous work, from his avant-garde poems and manifestoes to the trilogy of novels examined in the previous pages of this chapter, *: riflessi* (1908) *Il codice di Perelà* (1911) and *La piramide* (1926).[66] Thus, for example, in his brief introduction to the 1988 Mondadori edition, Fabrizio Bagatti writes that the *Interrogatorio* is a link between "la prima stagione del Palazzeschi narratore [...] e il periodo della maturità e del ritorno a un canone più classico di realismo romanzesco." He adds that the novel is a "nuovo e altrettanto rivoluzionario 'codice di Perelà'" and that there are also "chiari punti di contatto con l'esercizio di stile de *La piramide*." Luigi Baldacci's almost simultaneous yet longer essay is mostly devoted to clarifying when the novel was written, and it ends by discussing the text's "errori di stampa" and variants. More recently, Laura Lepri provides a more comprehensive reading, also reasserting that the Countess "incarna ancora una figura dell'evanescenza, un contenuto psichico ossessivo che la imparenta all'omino di fumo" (145), while Gino Tellini, in his long introduction to the Meridiani edition of

65 About the non-publication of the book, Laura Lepri notes that: "La sua struttura scissa, sperimentale, carente di progetto narrativo sistematico avrebbe rischiato di non essere accettata dalle convenzioni letterarie già impegnate nella discussione sul romanzo; e molto verosimilmente il personaggio, la sua 'morale', sarebbero stati fraintesi dal pubblico." Laura Lepri. *Il funambolo incosciente: Aldo Palazzeschi 1905–1914* (Firenze: Olschki, 1991), 146. At the same time, Palazzeschi's decision to not release a fully completed novel whose imminent edition had already been advertised at the end of the 1926 volume of *La piramide* edited by Vallecchi, could perhaps be interpreted as the latest "retentive" and implicitly subversive and defying act of a "letterato" who, similar to his fictional character (the Contessa), refuses to obey to the rules by which the dominant discourse/power (be it the publisher or the Italian literary world of the time) operates.

66 See Lepri's book-chapter "Il fantasma della natura" (137–47). Luigi Baldacci, in his "Interrogativi sulla Contessa Maria," *Paragone*, 1988, n. 458, 3–16, notes that *L'interrogatorio* and *La Piramide* share "soprattutto la tematica omosessuale" (10).

Palazzeschi's novels, traces in the text the "disinibita gaiezza che sprizza dal sottofondo tragico del manifesto 'Il controdolore.'"[67]

Taking the cue from the Countess's main trait, that is her being a "creatura golosa […] ingorda della vita […] ghiotta di sesso" (Tellini lxxix) and the abundance of alimentary references throughout the text, my objective in this section is to try to interrogate Palazzeschi's novel in the light of both the notion of "voraciousness," with its obvious alimentary declination and implicit link to the category of consumption, and Rita Felski's invitation, at the conclusion of her *The Gender of Modernity* (1995), to rethink "the history of the modern […] in terms of the various subaltern identities that have contributed to its formation."[68] A gendered approach to Palazzeschi may have some limitations and may raise some eyebrows, especially if one agrees, as I do, with Tellini's insightful observation that the two protagonists engaged in conversation in the novel, the sedentary (male) poet and the hyperactive Contessa, are actually the same person, "giusto il prisma delle antitetiche proiezioni dell'io che movimenta l'architettura compositiva de *La piramide*" (lxxxiii). And just to underline from an alimentary perspective this coincidence of identities, one could already observe that when the poet confesses that "le masticavo dietro un 'arrivederci' e non altro" (5); when, remembering the Contessa, he is affected by "una certa uggiolina alla gola"; or when he reveals that "nutrimento della nostra amicizia fu in gran parte la mia curiosità verso di lei […] amicizia [che] di altro sapore fra noi non avrebbe potuto durare" (10), the same language he uses betrays his own "hunger" as well, although the food he needs – "un bisogno naturale di verità" (3) – is apparently of a very different nature than the Contessa's.

The *Interrogatorio* is a text in which a male writer partially appropriates a feminine position to communicate his own desecrating criticism of the Italian cultural tradition, arguably using the feminine trope, as Andreas Huyssen has suggested, (also) "as a metaphorical threat to twentieth

67 Gino Tellini in: Aldo Palazzeschi, *Tutti i romanzi* (Milano: Mondadori, 2005), lxxix.

68 Rita Felski, *The Gender of Modernity* (Cambridge, MA: Harvard University Press, 1995), 212.

century political, social and linguistic structures."[69] It is also a text in which men's main reason of existence is apparently that of being "pietanze" or "contorn[i] della vita," "balocc[hi]" available to be consumed and to satisfy the Contessa's insatiable sexual appetite (56), and which is loaded with alimentary references. Given these features, it seems particularly appropriate, among the other novels in Palazzeschi's *oeuvre*, to be approached through the double critical lens of gender and food studies.[70] I would thus like to read this novel not only as another example of Palazzeschi's irreverence towards literary tradition and poetic inventiveness, but also within the wider framework of gender issues and a discourse which acknowledges the pervasiveness of food and alimentary imagery in avant-garde and modernist literature in general, one that in its alleged capacity to "elucidat[e] [...] the subject-object relation" (Delville 19) hopefully has, by now, already shown some of its critical versatility and hermeneutical usefulness.

If it is true that Palazzeschi's representation of a female voracious consumer in *Interrogatorio* constitutes one more example of the ways "femininity and modernity have been brought into conjunction" (Felski 9), addressing the specific characteristics and dynamics of such conjunction may provide further insights both on Palazzeschi's poetics in general, and on his own (potentially competing) attitudes towards modernity and femininity in particular. With this I mean that the Contessa, the character who articulates the author's own alternative, demystifying discourse about literature, life, and art in modern times, not only merges femininity and masculinity, but represents the combination of two cultural and ideological "moments." On the one hand, she seems to embody "everything that modernity was not, the living antithesis of the ironic self-estrangement of

69 Qtd. in Alyce Mahon, "Hans Bellmer's Libidinal Politics" in Raymond Spiteri, Donald LaCoss, eds, *Surrealism, Politics and Culture* (Burlington: Ashgate, 2003), 248.

70 Once the poet is given the label of "contorno" by the Contessa, he keeps the alimentary metaphor alive in his reply: "Un po' di cavolo ... strascicato, si e no, nella padella, brava. Però finite per confessare che se da una parte sentite il bisogno di quelli là, della vostra pietanza, non vi dispiace poi, dall'altro lato passare molte ore amichevolmente con me, e un po' più su nel discorso, dandole il suo bravo contorno" (56).

urban man," a "blissful and non-alienated" creature (Felski 50/56) who, by following her animal instincts, desires, and libidinal excesses, is as removed as possible from the commoditization, objectification, and rationalization usually associated with the sphere of a masculine modernity. On the other, however, she is the figure of an active and desiring female consumer, a self-conscious and "triumphant erotic subject" (Felski 206), the representative of a nascent and always more femininity-centered consumerist society and, as such, of a more comprehensive view of modernity that does not exclude "the realm of irrationality, aesthetics, and libidinal excess" (Felski 30). Building on and adapting to the circumstances some of Felski's insights, I would like to argue that Palazzeschi's recourse to alimentary imagery plays an important role in shaping this complex portrait of the Contessa and, by extension, in defining the writer's own interpretation and critique of modernity (and femininity) along a path that lies (or, perhaps better, "balances") between "dissacrazione" and "rientro nei ranghi" (Lepri 142).

To put it differently, if one "hungry half" of the Contessa resembles a sort of erotic-mythic reincarnation of a noble savage[71] and may implicitly suggest Palazzeschi's syntony and likely acquaintance with the thought of Georg Simmel, the German sociologist, who, addressing the issue of women's role in modern society, concluded that "femininity and modernity are irrevocably opposed" (qtd. in Felski 47), her other hungry, sexually voracious and consumerist half seems also to simultaneously reflect the antithetical conclusions of Otto Weiniger, the Austrian philosopher, who, in his widely influential *Sex and Character* (1903), viewed instead "modernization [as] synonymous [...] with an inexorable and threatening process of feminization" (qtd. in Felski 46–7).[72]

71 The following quotes reinforce this association with the "savage": "E scattata in piedi, si dette a passeggiare come una fiera nella gabbia" (54); "La sensualità deve venire dai muscoli, nutrita da essi, non dai nervi o dal cervello" (67).

72 The work of both Georg Simmel (1858–1918) (i.e., *On Women, Sexuality and Love*) and Otto Weiniger (1880–1903) was widely known and influential among European writers and intellectuals active in the *fin de siècle* and in the first decades of the twentieth century. Interestingly, in his essay *Sociology of the Meal*, Simmel observed that "The shared meal elevates an event of physiological primitiveness [...] into the

The fact that the Contessa embodies these two opposite and yet interconnected positions tangentially confirms, from yet another perspective, Tellini's point that the poet and the Contessa are two faces of the same medal: she is a woman and yet often thinks and behaves like a man. A descriptive statement such as "vestiva con semplicità squisita concedendosi spesso un leggero tono così garbatamente mascolino ch'era in fondo il suo più bel capriccio femminile" (8) or a reference to her style as being "di una spregiudicatezza virile ed insieme della più femminile vivacità" (13) easily provide further textual evidence of this situation. However, perhaps even more interestingly, at the level of genealogies, this general hybridity (if not actual androgyny) of the Contessa may in turn suggest that also "Equilibrio," the third among Palazzeschi's manifestoes published in *Lacerba* (III, n. 4 1915), is among the sources of the *Interrogatorio*, in addition to the more often cited "Controdolore." As Anthony Tamburri reminds us, "'equilibrio' equals social conditioning [...] Whereas for society 'equlibrio' is adherence to the conventions of an already existing state, for Palazzeschi it consists in (1) the ability to adjust and acclimate oneself to opposing and divergent influences, and (2) a strong desire and capability to explore and experience these different forces" (Tamburri 2003, 52–3). The Contessa's hybrid nature between the "opposing and divergent influences" of femininity and masculinity, her overall unconventionality and irreverence towards any social conditioning and her being simultaneously primitive and modern, make her the perfect incarnation of the vital, inherently democratic life-philosophy presented in "Equilibrio," a philosophy, which, as I previously noted, Palazzeschi summarizes with a reference to food and eating at the conclusion of his manifesto: "Si è detto che il mondo è come un'arancia. Giustissimo. E tu ti contenti di mangiartene uno spicchio? Cucù! A me fa gola tutta!"[73] Expectedly, since this world/orange is very tempting for the

sphere of social interaction," implicitly pointing to the distinction between food as matter (primitive) vs. food as consumption (modern) (in *Simmel on Culture: Selected Writings*, David Frisby and Mike Weathersone, eds (Sage Publications: Thousand Oaks, CA. 2000), 130.

73 Of course, in the traditional sense of the word, one can find "equilibrio" also at the formal/structural level of the novel, with its two sections of similar length (dialogical in the first and first-person narration in the second).

Contessa as well, the same notion of "equilibrio" makes repeated appearances in her various lucubrations:

> Nulla mi sfugge di quello che si dice o si pensa di me da tutte queste rispettabili persone [...] e mi serve per poter misurare la loro imbecillaggine o ipocrisia [...] il loro disagio difronte al mio *equilibrio* (23); [...] una donna, non cosidetta, ma vera, degna di portar questo nome, piace [...] forza risultante dall'*equilibrio* di tutte le facoltà fisiche e morali nel loro sano e giusto sviluppo (53); molte di quelle bocche d'ignoranti e di sciocchi, rivelano sovente [...] un *equilibrio* [...] cui non arriveranno mai i tuoi cari cervelloni (55) [italics mine]

The last statement of the Contessa, according to which such "equilibrio" is more often found at the level of some "fools' mouths" than at that of some "brainy intellectuals," not only alludes to the situation of a degraded and conventional [modern] culture, one just able to "difendere e [...] trasmettere la mistificazione" (Lepri 140), but it belongs to and evokes again, through the reference to the "mouths," the alimentary imagery in the novel.

At the more superficial level, one might say that such imagery is just one more reflection of Palazzeschi's colorful style, linguistic playfulness, and general reaction towards the selective, "high" literariness of the traditional (patriarchal) literary establishment, represented by those writers, from Leopardi to Carducci to D'Annunzio, whom the Contessa disparages right before beginning to tell the story of her life in the second half of the book. This is what one may observe, for example, when the poet replies with the expression "Trippa da gatti" to the Contessa's description of history as a "grande pentolone dove ci bollono dentro tutti i romanzi" (41); when the two characters engage in entertaining, sexually allusive conversations about the amount of chicken one is able to eat at different stages of one's life ("A venti anni si mangerebbe un pollo a cena, non è vero? o a colazione, a quaranta la metà sì e no, a sessanta basta un quarto, a ottanta un caffè e latte sembra troppo" [20]); or when they argue about the different ways one may prepare a specific course ("Certe pietanze voi le potete cucinare in cento modi, lo so, contornare, ricoprire, accompagnare con cento salse, ma sotto sotto ci sarà quella pietanza, mi capite?" [36]). In short, this recurrence of alimentary references may be initially explained by stylistic justifications and the need to lower the linguistic register to the earthly and passionate physicality of the protagonist.

At a second level, however, the various instances in the *Interrogatorio* in which the Contessa, in her role as a quintessential "consuming subject," is associated with food and consumption not only remind us of her voracious sexual behavior, but also contribute to situate "femininity at the heart of the modern in a way that the discourses of production and rationalization [...] did not" (Felski 61). In his depiction of the poet's encounter with the Contessa and then of her self-narrated evolution from young member of an ancient but decayed aristocratic family to independent and free-thinking woman of "vituperevoli costumi" (15), Palazzeschi, on the one hand, denounces the actual misperception and consumption in modern times of the qualities embodied by the Contessa ("la sincerità, la trasparenza, la spontaneità" Lepri 138), but also implicitly alludes to the rise of a new cultural and economic era, one in which society as a whole is "increasingly structured on the imperative to consume" (Felski 78).

Similar, at least in terms of her sexual profligacy, to Zola's Nana, the protagonist of *Au bonheur des dames* (1883) [The Ladies' Paradise] analyzed by Felski, one could argue that this "all-consuming side" of the Contessa seems to be "constructed through modernity" as well. Although originally coming from a "città piccola e provinciale, un guscio d'uovo affumicato," she certainly belongs to and thrives in modern urban culture. Perceived by conventional, bourgeois society and morality as a prostitute, an ex dancer or a spy, she nonchalantly disregards public opinion and remains for most people a mysterious and enigmatic presence. Her constant smoking and her use of the busiest "caffè" in Florence as hunting ground (a place where besides sipping coffee, she can simultaneously prey on men), immediately associate her with the figure of a *femme fatale*, that is, someone who, by definition, incarnates the potentially destabilizing and threatening (even deadly) aspects of uncontrolled female desire and sexuality. Her understanding of life as "action" and, later, her describing it in purely economic, "squandering" terms, is in line with the new ethos of consumerism and antipodal to the bourgeois world of temperance and self-restraint (see Felski 78): "la vita è azione, figliolo caro, e starsene col sedere sulla seggiola non è azione [...] la vita si vive a contanti e si liquida giorno per giorno, ora per ora" (10/70). Finally, her omnivorous attitude erases traditional class differences by indiscriminately accompanying herself and, in turn, "feeding" with men of all different sorts, "Ufficiali, sottufficiali, soldati, studenti, un vetturino colla *tuba* in testa, un meccanico colla *tuta*, un garzone di macellaio, un pompiere, uno *chaffeur*, una guardia comunale, un carabiniere, un finanziere, un ferroviere" (16).

It is not surprising that this passage echoes a similar one in "Equilibrio" in which Palazzeschi praises difference and variety in culinary terms: "Voi cercate di fornire al vostro palato, al vostro stomaco, sapori e cibi differenti. Dopo aver mangiato un pezzo di bue sanguinante vi piace mettervi dentro della fresca insalatina, e fate benone" (30).

To be fair, it must also be noted, however, that despite this rich and varied diet, this *mangeuse d'hommes* does not consider herself a prostitute at all ("Io detesto la prostitute" [23]) and refuses any compensation for what she does. When one of her lovers feels obliged to "corrisponder[le] la mercede" (33), she tells him: "io non mi sono fatta mai pagare da nessuno [...] Quello che fo, lo fo così ... per sport; tieniti pure le tue cinque lire" (34). This statement not only reminds us that her activity – precisely like the art she says to despise ("sono un capolavoro della vita non dell'arte") – is fundamentally unproductive and wasteful, and thus even more uncomfortable for the same modern, protocapitalist society she nonetheless belongs to and embodies, but also that, again resembling Nana, "her contempt for money is simultaneously an expression of disdain for the entire system of cultural values premised on the assumed authority and prestige of traditional symbols of masculinity" (Felski 78).[74]

74 Laurie Teal's examination of the changing perceptions of the figure of the prostitute during different stages of the modernist period has also been a source of inspiration for this section of my discussion. As Teal notes: "The prostitute (the Weiss), as a newly constructed 'ideal' of femininity-as-flesh, stands in opposition to the artificially adorned femininity represented by commodity culture, and the Weiss thus rather confusingly becomes an ally in modernism's campaign against bourgeois femininity (Weisslichkeit) [...] Kraus's aphorism ironically places the prostitute, in this early stage of modernism, outside the corruptive forces of capitalism, outside of culture and society altogether. In a world of cheapness, imitation, and illusion produced

This ambivalent nature of the Contessa becomes more evident if one notes that while her appetite for life places her "at the heart of the modern," making her an icon of and in line with the consumptive tendencies of an expanding consumerist, capitalist society, precisely that same discourse on voraciousness and hunger cannot but also "underscore the animalistic and instinctive nature of female appetite" (Felski 76). The alimentary imagery, in other words, simultaneously contributes to construct the Contessa's antimodern and primitive persona and thus paradoxically situates her "outside the corruptive forces of capitalism, outside of culture and society altogether" (Teal 80).

In several instances, and in spite of the ironic and parodic register of the text, the Contessa's association with food seems to simultaneously emphasize her antimodern traits. When, for example, the poet notes that the Contessa "divorava una dopo l'altra [...] parecchie diecine di sigarette" (5), or when he replies to one among the Contessa's most provoking affirmations: "un'ora di vita è sempre, in ogni caso, da preferirsi a un'ora d'arte," with a snappy "cosa siete, morta, quando gustate un capolavoro?" (51), his lexical choices are particularly revealing. A verb such as "divorare" applied to cigarettes, for example, while aptly indicating the extraordinary amount of smoke inhaled by the Contessa, has also the effect of not only underscoring the transgressive nature of the gesture for a woman, but also of transforming an activity (smoking) which, at the time, could still be associated with ideas of culture, modern manners, class and social refinement, to the animalistic satisfaction of an excessive desire. A similar dynamics seems to

---

by expanding capitalism, the prostitute represents to the modernist an apparent holdout of the 'real,' a place to encounter bodily experience at its most immediate. Robert Scholes suggests that 'the brothel of modernism was also a sanctuary, a relic of a time when the flesh at least counted for something' (204). Toward the turn of the century, however, as sexuality and desire become increasingly mediated by the cerebral investment of looking in commodity culture (advertisements, display windows, exhibitions, catalogs, fashion shows, etc.) both the streetwalker and the brothel prostitute begin to place themselves (both consciously and unconsciously) in the same position as other objects on display for sale." Laurie Teal, "The Hollow Women: Modernism, the Prostitute, and Commodity Aesthetics," in *Differences: A Journal of Feminist Cultural Studies* 7.3 (1995), 80–1.

characterize also the second quotation, in which the poet's choice of the verb "gustare" to express the Contessa's enjoyment of an artistic "capolavoro" reduces what is supposed to be an intellectual, cerebral practice to the mere satisfaction of a primal appetite.

Along the same lines, when the two characters decide to attend *Norma*, Bellini's opera,[75] the Contessa's unwillingness to refresh her spirit ("un gradito ristoro") (47) without, at the same time, providing a very different kind of refreshment to her body with the "caffettiere e aranciaio della galleria" (48), is that which first causes a reaction of disgust in the poet ("Vi alzate ... e ve ne andate [...] è proprio quello che mi disgusta in voi, che mi dispiace, è una questione di forma [...] Mi fate vomitare" [49/50]), and then prompts him to explicitly juxtapose his own cultured, rational modernity (according to which either music or "qualsivoglia opera d'arte" should be enjoyed and consumed "con profitto" [47)] with the Contessa's natural, impulsive (and, thus nonprofitable, "squandering") "primitiveness": "Io non sono un uomo della natura, coi vostri sistemi ci riportate ai tempi di Adamo ed Eva, nel paradiso terrestre. In tanti secoli la civiltà ha camminato" (51).

Despite the poet's disgust and impulse to throw up, there are few doubts that Palazzeschi voices through the Contessa his own complaints and nostalgia about the loss of the "natural" and the lack of authenticity in and perversion of modern culture and art. His appropriation of the category of femininity to articulate his counter-discourse therefore, is most likely not aimed at strengthening his authority as male writer but, rather, to criticize values and behaviors generally "associated with the ideology of bourgeois masculinity" and to express a "sense of alienation from dominant social structures and his own class identity" (Felski 101). Similarly, one could safely add that the *Interrogatorio* displays the rejection of "dominant views of female passionless, passivity, and conservatism, [ultimately] present[ing] woman as a modern, self-conscious, and triumphant erotic subject" (Felski 202).

75 The Contessa will behave in the same way when attending *Il Barbiere di Siviglia*. Ironically, *Norma*'s most famous aria is "Casta Diva."

This said, after having pointed to the role of the alimentary discourse in "balancing" the Contessa's situation both inside and outside the modern, one may still wonder, especially in light of the events in the second half of the novel and its conclusion, if the picture I just delineated may be complicated a little. This section of the *Interrogatorio*, in fact, is where the Contessa becomes the protagonist of a melodramatic and parodic *feuilleton* that tells the story of her life. We thus learn of her isolated, aristocratic childhood, her precocious and active discovery of the other sex (Ranieri), her escape to Rome in search of freedom and independence, her return to and segregation in the family palace together with a comical wet nurse ("il tacchino") who likes drinking, her fake repentance and, eventually, her definitive escape to Paris with her new lover, Beppino, a young officer she had managed to meet by easily corrupting her guardian with a good bottle of wine. As Lepri summarizes, Palazzeschi here "rilegge personaggi e moduli del romanzo per signorine smascherandone la funzione di trasmettitore di contenuti convenzionali" (Lepri 141). References to the alimentary sphere abound also in these pages, from those alluding to the, by now, familiar voraciousness of the Contessa ("mi comportavo con tale naturalezza, forse già pregustando lontano qualcosa di più saporito che non fosse l'amore del mio compagno" [85], to her "indigestibility" for Ranieri (and men like him) ("Aveva mangiato per un anno intero una pietanza che non era ancora per il suo stomaco e gli era rimasta tutta lì, senza poterla digerire ... e aveva vomitato" [91]), to the dietary, *contrappasso*-like punishment she and her plump, wine-and-food-loving guardian have to suffer during her segregation: "Come c'entrano i cibi, colla sua punizione? Ella non ha peccato di ghiottoneria [...] Quello che ci portano su non è il pranzo della tavola padronale, o lo è solo in parte" [117]). At the conclusion of her story, then, the Contessa reasserts to the poet her independence and distinction from the standard, object-of-consumption-status of other women with a last culinary statement: "anche di me volevano fare una polpetta, non ci sono riusciti" (140). However, although it may be true that the Contessa managed to escape from her reduction to "edible matter" and, implicitly, from a primitive, old-fashioned perception of femininity, she does not seem to have escaped the fate of being paradoxically turned into art, as her tragicomic leading role in the autobiographical fiction within the fiction would seem to suggest. In other words, the dynamics in this section of the novel strongly resembles one by which "the threat of the natural" (or the "primitive"), that is the "excess of and unconscious and undisciplined subjectivity of a woman's body," is tamed and "negated by being turned into art" (Felski 110), thereby making it again more digestible and consumable by the male poet and his readers alike. By turning the woman who said that she was a "masterpiece of life and not art" into precisely a producer and a "work of art" herself, and then – in accordance to a characteristic conclusive pattern in Palazzeschi's fiction, from *: riflessi* to *Perelá* – eventually making her vanish, the writer contains the perceived menace her unconventional, uprooted, and nomadic subjectivity represents for his readers (and simultaneously, still manages to convey through the more traditional form of the parodic feuilleton her desecrating message).[76]

It seems hard to deny, however, that in the end the Contessa is placed out-of-the story (and history) altogether, since she disappears from the poet's sight, saying that she's going (among all places) to the stereotypically natural and still primordial Sardinia in search of yet another man, only to add immediately after that she may soon come back to modern culture and society: "può anche darsi che vi rimanga pochissimo in Sardegna, a seconda dei casi, e conto, al ritorno, di fermarmi a Roma un pezzetto" (147). As a matter of fact, it seems that whether the Contessa embodies a primitive woman or her opposite, an antipodal (too) modern "itinerant woman" always in search of new spaces (and foods), and bearer of a heterogeneous, obscene, and omnivorous diet and discourse that counters the stale, monotonous one of the (patriarchal) Italian literary tradition, her outrageous femininity and her story have an ambiguous, dazzling, and blinding physiological effect on the poet. The latter's reaction to the departure and vanishing of the Contessa: "fra la gente nel rumore del via vai" (147), in fact, may resemble that of someone who has just emerged from an enlightening and satisfying mystical experience, but also suggests

76 For the notions of "uprootedness" and "nomadism," I am referring to Rosi Braidotti's *Nomadic Subjects: Embodiment and Sexual Difference in Contemporary Feminist Theory* (New York: Columbia University Press, 1994).

his exposure to a debilitating, almost annihilating experience: "Caddi sulla poltrona di dove ella s'era allora levata, riverso cogli occhi erranti nel vuoto, che non cercavano più, che non vedevano più, e le braccia abbandonate: 'Che è? Che è? Che è stato? Che mi passò vicino? Che?'"

The conclusion of the *Interrogatorio* resonates with the poet's unanswered question. The reader, in the end, may still be wondering what kind of text has been served and – given my observations about the aesthetic "digestibility" of the Contessa – whether she did not speak too early when she stated that "anche di me volevano fare una polpetta, non ci sono riusciti." Finally, one may wonder if the fate of the never-released *Interrogatorio* could not be further indication of Palazzeschi's own "rientro nei ranghi." That is, his withholding of the novel may indicate his potential discomfort in releasing within the Italian literary landscape in general, and the male *milieu* of the Florentine avant-garde in the 1920s in particular, not only a text with such an "incendiary," voracious character but one with the specific portrait of a woman with the potential of shaking, if not literally eating away the very foundations of male literary culture.

CHAPTER THREE

Hard to Swallow: Two Fictions by Paola Masino and a Play by Massimo Bontempelli

3.1 Indigestible Fictions: Paola Masino's *Fame* and Massimo Bontempelli's *La fame*

> Da quando gli uomini adoperano la forchetta e fanno irriconoscibili le carni, non sono più scesi Iddii sul mondo
>
> — PAOLA MASINO, *Monte Ignoso*

One of the welcome effects of the recent expansion among specialists of Italian culture and literature in academic disciplines such as gender, cultural and women's studies is the rediscovery of modernist writer Paola Masino (1908–1989) and her unconventional literary production. Unavoidably overshadowed during her artistic career by the achievements of the two giants Massimo Bontempelli (1878–1960) and Luigi Pirandello (1867–1936),[1] Masino's peculiar position within the canon of modern Italian literature is beginning to emerge thanks to the efforts of a number of excellent scholars who have emphasized her originality.

Lucia Re, for example, in a study dealing with fascist theorizations of gender, notes how Masino's novels "clearly belong to a different and unusual experimental register, a modernist realism infused with surrealist

1 Bontempelli was her life-time companion and Pirandello a life-time friend.

and expressionist elements"[2] and, in an even more recent essay, focuses on the political relation between hunger, food, and antifascism in Masino's last novel *Nascita e morte della massaia* (which I will discuss in the next section of this chapter).[3] Fulvia Airoldi Namer, after pointing out Masino's "particolarissimo 'realismo magico' [...] nell'ambito del 'novecentismo,'" details a pattern of recurrent chtonic *topoi* and descending, visceral, feminine motifs in her novels.[4] Tristana Rorandelli, in turn, by addressing the themes of the female body and of maternity during fascism, convincingly demonstrates that *Nascita* subtly questions fascism's reductive and narrow conception of womanhood. Lastly, Allison Cooper confirms Masino's "radical transformation of traditional narrative elements [...] departure from the classic conventions of storytelling" and "challenge to traditional narrative structure," and suggests that her work (*Monte Ignoso*, in particular) "reacts to and critiques the post-war search for epistemological 'certainties' and the very notion of a return to order and tradition."[5]

2 Lucia Re. "Futurism and Fascism, 1914–1945" in Panizza, Letizia and Wood, Sharon, eds, *A History of Women's Writing in Italy* (Cambridge: Cambridge University Press, 2000), 190–204, (203).

3 Lucia Re. "Fame, cibo e antifascismo nella Massaia di Paola Masino" in Muzzarelli, Giuseppina – Re, Lucia. *Il cibo e le donne nella cultura e nella storia*. Prospettive interdisciplinary (Bologna: CLUEB, 2005), 165–81. Re interprets Masino's *Nascita* as an "audace atto di accusa contro il fascismo" and observes that "Tale atto trova nel campo metaforico del cibo e del nutrimento in relazione al corpo e al ruolo femminile il suo più efficace e suggestivo strumento" (Re 2005, 165). Re shares my interest in the alimentary motif in Masino's novel, but my analysis will purposely marginalize the political and the gender aspect and privilege instead the aesthetic/hermeneutical one.

4 Airoldi Namer, Fulvia. "La terra e la discesa: l'immaginario di Paola Masino", *Otto-Novecento*, vol. 24, n. 3 (September–December 2000), 161–86, (165).

5 Allison Cooper, "Gender, Identity and the Return to Order in the Early Works of Paola Masino," in (L. Somigli and M. Moroni 2004, 378–99), 379/384. For additional, recent criticism on Masino see also: Paola Blelloch, "*From Trunk to Grave: the Hallucinated Story of a Housewife*," *Nemla Italian Studies*, 13–14 (1989–1990), 89–103; Manuela Gieri, "*Paola Masino*" in Luca Somigli and Rocco Capozzi, eds, *Italian Prose Writers, 1900–1945*. (Detroit, MI: Thomson Gale, 2002), 193–9; Silvana Cirillo, *Nei dintorni del surrealismo, Da Alvaro a Zavattini umoristi balordi e sognatori*

This quick glance at the most recent critical essays devoted to Masino's fiction reveals the variety of approaches it lends itself to, its multifaceted complexity, and its relevance in providing a more complete picture of a modernist discourse (both Italian and European) that has too often limited itself to considering a fixed list of canonical male authors.

This said, the question of the intellectual exchange that took place during the life-long relationship between Paola Masino and Massimo Bontempelli – a prominent and "canonical" author for sure – and, in turn, the presence of thematic and ideological analogies between some of their respective works has been rarely addressed in detail by scholars of Italian modernist literature.[6] Dated and imprecise attempts to evaluate Masino's debt to her older and more established companion or to assess whether her striking writing style was influenced by Bontempelli's "realismo magico," have been replaced recently by the realization – supported especially by the study of the intense correspondence between the two writers and with other figures – that this influence was not one-sided. As, for example, Marinella Mascia Galateria observes, if, on the one hand, "nella scrittura di *Monte Ignoso* entrano anche i consigli di Bontempelli [...]" it is, on the other, "l'influsso di Paola Masino che gli suggerisce le fabule di *Vita e morte di Adria* e di *Gente nel tempo*" and that "insinua e [...] accampa l'idea della morte" in Bontempelli's fiction and theatrical production.[7] Fulvia Airoldi Namer, in turn, notes that the issue of the similarities in the fiction works of Masino and Bontempelli "è un tema che meriterebbe uno sviluppo a parte," implicitly leaving the undertaking of such a "sviluppo" to future scholars (Airoldi Namer 2000, 182n).

---

nella letteratura italiana del novecento (Roma: Editori Riuniti, 2006) (there is a chapter on Masino as a writer of the "fantastic").

6 See, for example, besides the cited works of Fulvia Airoldi Namer and Marinella Mascia Galateria: Louise Rozier, *Il mito e l'allegoria nella narrativa di Paola Masino* (Edwin Mellen Press, 2004); Sharon Wood, "Feminist Writing in the Twentieth Century" in Peter Bondanella and Andrea Ciccarelli, eds, *The Cambridge Companion to the Italian Novel* (New York: Cambridge University Press, 2003).

7 Marinella Mascia Galateria, "L'autobiografia trasfigurata di Paola Masino", *Avanguardia*, VI, 17 (2001), 110–31.

Following up on Airoldi Namer's recommendation, I intend to consider in this essay two works: Masino's short story *Fame* (1933) and Bontempelli's almost homonymous three-act theatrical *pièce La Fame* (1934). These two works are fertile ground on which to discuss the alleged reciprocal influence of these writers on each other and, in particular, to address the ramifications of the alimentary rhetoric which pervades both texts.

The scant criticism on either Masino's tale or Bontempelli's play has generally underlined the transgressive socio-political message contained in each, as they expose the failures and weaknesses of fascist ideology, demystifying its propaganda and alimentary policy and, as a consequence, the parallel episodes of censorship that followed their appearance on the Italian cultural scene.

Maria Vittoria Vittori, for example, in her introduction to Masino's short story collection *Colloquio di notte*, mentions first Mussolini's suppression of the magazine *Grandi firme* and then notes that among the privileged themes in *Fame* is that of "paternità [...] una paternità così sofferta nella carne e nello spirito da non negarsi al desiderio di autodistruzione espresso dai figli" (in Masino 1994, 27). Airoldi Namer's contribution reiterates the point while spottily summarizing its storyline: "*Fame* era stata pubblicata nel 1933 in una piccola rivista senza suscitare critiche. Fece scandalo quando nel 1938 Zavattini la pubblicò su *Grandi Firme*: la rivista stessa fu soppressa. È la storia di tre [actually, they are two] bambini affamati che chiedono al padre di ucciderli. Egli li strangola, poi va a costituirsi: racconterà tutta la storia in cambio di un piatto di minestra" (Airoldi Namer 2000, 177n). Finally, Lucia Re reminds us that Masino's dramatic tale of poverty, death, and starvation was a practical denunciation of the "fallimento del grande, spettacolare sforzo della cosiddetta 'battaglia del grano' lanciata da Mussolini per sfamare la nazione" (Re 2005, 178).

As far as Bontempelli's *La fame* is concerned, Silvio D'Amico's 1949 critical review for the journal *Sipario* stands out for its brevity and, as far as I have been able to establish, its singleness.[8] After mentioning the fascist

8 Silvio D'Amico, "The Italian Drama in the Past Five Years," English translation by H. Clancy, *Educational Theatre Journal*, vol. 2, n. 3 (October 1950), 208–17. D'Amico

"prohibition of the representation" of the play in 1936,[9] he also offers a useful, although not entirely accurate, summary of its plot:

> The story of a woman [Barbara] who, after suffering unspeakable hardships during an enforced exile, comes back to her home town to carry the word of peace to her fellow citizens who are divided into two sternly opposed factions [...] when, after forgiving those who had tortured her in many ways, she meets the man who had submitted her to the most atrocious torture – hunger – she revolts and renounces her former plea for peace. There will be no peace in this world, she says, as long as hunger exists (213).

D'Amico concludes by minimizing the politically subversive content of the play ("the author only wishes to express a lyrical, human cry of pain and resentment") and commenting negatively on the efficacy of its dramatic representation ("characters [...] are only geometrical and arid symbols [...] they have no humanity").

Given the closeness of Masino and Bontempelli in these years, the choice and pervasiveness of the alimentary motif in both of these works from the early Thirties suggests a silent, between-the-lines conversation between them. One can imagine an ongoing dialogue on the topic of hunger or, perhaps, a sort of modernist *tenzone* between two writers who were, each in her/his own way, artistically confronting both a theme and a practical problem ("la fame") that was particularly cogent and felt in fascist Italy. Furthermore, poverty and hunger were issues that both writers had experienced in their own lives,[10] as the passage below in the section "Gli

---

was a famous theatre critic who created the Accademia nazionale d'arte drammatica in 1935.

9 Bontempelli explicitly manifested his rebellion against the regime in at least three, well-known occasions. First, with his play *La Fame* (1934); second with his commemorative discourse on D'Annunzio (1938) and, third, with his refusal to occupy the position of professor of Italian Literature at the University of Florence, a position which would have required him to take the place of Attilio Momigliano, who had been affected by the racial laws (see Airoldi Namer).

10 It is interesting to remember that F.T. Marinetti's and Fillia's *La cucina futurista* made its appearance in the same year, 1932. Obviously, Marinetti's text dealt more

anni parigini" (1930) in Masino's autobiographical book, *Io, Massimo e gli altri*, clearly demonstrates:

> La povertà mia e di Massimo era di una specie particolare [...] che ci consentiva di saltare un pasto e comprare un fiore senza dubbi e senza soffrire la fame [....] Spendevamo come ci veniva in mente, certi che domani o fra un'ora qualche cosa sarebbe accaduto per la quale avremmo potuto ancora mangiare benché per il momento non avessimo un soldo in tasca (Masino 1995, 33–4).

Because of the multifaceted connotations and the symbolism present in the hunger motif, this shared, literary interest in the topic of hunger may also allude to "problems and themes" that refer more specifically to the aesthetic field and to questions of literary representation. Lucia Re, for example, noted that the motif of hunger explored by the writer in her short story will eventually be associated with a subversive and scandalous form of art itself, one linked to the idea of a refusal of a certain socially and culturally prescribed female identity, a theme repeated in her most famous novel, *Nascita e morte della massaia*, written a few years later.[11]

These two works, therefore, offer the opportunity to reflect on the creative relationship and gender-influenced poetics of two modernist artists who were intimately, and yet differently, dealing with issues of representation, expression, and creativity – not to mention with their own ambitions and subtly antagonistic appetite for affirmation and recognition in the Italian literary and publishing landscape of the period.[12] Once again, another passage marked by several references to food and eating in

---

with the utopian ways one should (or should not) eat than with the tragic effects of a real lack of food.

11 Writing on Masino's novel *Nascita e morte della massaia*, Lucia Re notes that: "La fame della protagonista è infatti la sua vera e sola 'arte', e sta a significare innanzitutto proprio il rifiuto di diventare il tipo di individuo e soprattutto di donna che il fascismo e l'ideologia borghese e patriarcale vorrebbero" (Re 2005, 179–80).

12 It may be worth remembering that Pirandello's novel *Suo marito*, narrating the vicissitudes of a gifted, young female writer (Grazia Deledda) and her ambitious husband in the Roman literary world of the beginning of the century, appeared in 1911.

Io, Massimo e gli altri seems particularly revealing in the light of the previous considerations:

> Dove andare oramai per trovare qualche cosa d'improvviso, di eccezionale? Lo sconforto abitua ad accontentarsi e questo è il peggio [...] Mi pare che di questi tempi la gente stia intorno alla tavola piena d'avanzi del passato e con uno stuzzicadenti chi si pulisce la bocca, chi scherza con un ossicino, chi fa palline con le briciole di pane, qualcuno perfino pilucca muffiti chicchi d'uva, e altri ritenendosi coraggiosissimi sovvertitori di ogni ordine stabilito, rovesciano qua e là il sale sulla tovaglia unta. Credono di banchettare, immaginano di fare delle orge: sono anni che non mangiano, ma rimangono là intorno seduti, convinti di essere tanto sazi di cibi e di vini da non avere più la forza d'alzarsi. È inedia. Tuttavia scherniscono chi davanti a loro, seduto in terra in un angolino pulito mangia un bel pezzo di pane appena cotto. Si credono signori e puzzano d'unto, si credono rigurgitanti di nutrimento e sono più affamati degli attori che si affannano intorno ai polli di cartone, si credono generosi e sono strozzini, previdenti perché non giudicano e sono disertori. Insomma vedo davanti a me uno di quei quadri di Brueghel il Vecchio, dove a mezzo di tutta l'umanità sono rappresentati i sette vizi capitali. Ma c'è un vizio che allora credo non si conoscesse ed è quello del plauso perenne (Masino 1995, 75–6).

After having mentioned the poor reception that the texts under examination had at the time of their appearance, as well as their unconventionality and shock-value, Masino's frustrated search for a different kind of aesthetic experience seems worth noting. Her desire for an artistic artifact with characteristics of "unexpectedness" and "exceptionality," one able to interrupt the self-congratulatory mediocrity of the status quo, could be interpreted as an invitation to the public and the critics to consider her own and Bontempelli's more avant-garde artistic production as exemplary answers to her initial question.

However, – Masino notes – most people (art and literature critics, intellectual readers) are apparently still contented to feed themselves with "left-overs" from the past. They feel satiated and believe in the relevance of playing with "old food scraps" or, out of the allegory, with a *passéiste* artistic cultural production, rather than looking for novelty and accepting differences. As a result, such readers remain essentially "hungry" and "starving." They would actually be eager to eat and enjoy something fresh and new, like a nice, just-baked "piece of bread," and yet, at the same time, they are also tragically unable to understand and judge its taste and quality. Basically, they seem to have no idea of who they are and of what their role should be; their own existence is fed only by misconceptions of themselves and of others. As she addresses the stagnant condition of modern art, the blind and self-aggrandizing attitudes of criticism and the subsequent difficulties for another kind of still marginalized artist/producer to emerge and thrive in a particular cultural environment, Masino's recourse to a peculiar allegorical, alimentary rhetoric may have understandably captured the reader's attention in the previous passage.

A similar juxtaposition of the themes of cannibalism, artistic creation, and parenthood recurs in a letter to her parents dated February 3rd, 1938. In this letter, after having observed that "l'unico vero talismano per dedicarsi a qualunque arte [è] quello di non avere legami e pensieri materiali di nessun genere," thus alluding to the ascetic quality of the artistic process, and that "è meglio un'intelligenza un po' confusa magari, ma spregiudicata, che un buon senso tradizionale, anche per fare e allevare i bambini," she concludes by remarking that: "forse è bene che io non abbia bambini, se no li ridurrei cannibali. È meraviglioso che io abbia voi per babbo e mamma, se no cannibale mi sarei fatta per natura" (Masino 1995, 65–6). If my interpretation of these words is correct, their meaning is that Paola is glad not to have children because she might have been an absent, bad mother, unable to feed and take good care of them. If she had become a mother, she fears her children would have ended up feeding on each other and even on their own mother ("li ridurrei cannibali").[13] Masino then expresses gratitude for her own parents because, arguably, they never pushed their artistic daughter to have children, as is usual with other "normal" parents, and they let her follow her creative aspirations. If she had begotten children, she believed, she would have "naturally" become herself a "cannibal," suggesting that she would have joined the list of fictional female child murderers by figuratively eliminating and ingesting them.

13 Annie Vivanti's novel *I divoratori* [*The Devourers*] (1911) comes also to mind at this regard.

These lines are interesting for several reasons, not the least because they undoubtedly confirm Masino's anomaly and eccentricity among Italian women writers and intellectuals of her generation. However, given the specific alimentary context I am dealing with, they resonate because they address, through the confessional and more realistic epistolary format, some of the issues *Fame* had dealt with a few years earlier in fictional terms. Both the short story and the letter, in fact, contain more or less explicit references to the kind of sacrifices, costs, and renunciations necessary to create and "make something grow," be they works of art or children, and they also suggest the possibility of a link between the theme of hunger (a "pensiero materiale"), on the one hand, and that of infanticide (i.e., the allusion to cannibalism and some form of abortive nurturing) on the other. The next section of this essay will investigate in more detail the presence and the ramifications of this crucial relationship first in Masino's short story and then in Bontempelli's play.

The key question in *Fame*, especially in the light of Masino's epistolary comments on creation, motherhood, and cannibalism, seems to deal with the individuation of the cultural and literary implications associated with infanticide and, ultimately, with the meaning of her selection of a male protagonist, a father, as the perpetrator of this highly symbolic, taboo-act. Lucia Re, although not directly addressing *Fame* but, instead, historically contextualizing Masino's 1939 novel *Nascita e morte della massaia*, observes that the theme of hunger in this work is linked to two other issues about which fascism was very sensitive: war and abortion. Since fascism considered abortion an offence against race and religion, to justify it, as the *massaia* protagonist in the novel implicitly does, contests the idea of woman theorized by the most extreme form of fascism, which reductively envisioned a woman as someone who needs to "maturare nel suo corpo [e] nutrire il figlio" (Re 2000, 180n), someone who has to feed herself in order to give food to her children. The combination hunger/abortion, therefore, achieves in the novel a clear, transgressive "symbolic resistance" both against fascism and the established standard of sexual roles (Re 2000, 180). Mary Kay Miller, in her book-chapter entitled "Writing Infanticide", adds that as a theme and literary figure, infanticide "speaks violently and eloquently of the conflict between women's traditional role as reproducers and their desire to fashion a productive literary role" and that among a number of female writers belonging both to Western and Non-Western literary traditions, the function and role of "reproduction is not categorically rejected [...] but rather reconfigured and re-presented in such a way as to free the feminine from a cycle of imitation and perpetuation."[14]

Although not specifically referring to a modernist Italian literary *milieu*, Miller notes that infanticide appearing in women's texts "can sometimes be read as a metaphor of liberation, a freedom from reproduction [...] [it] constitute[s] a challenge to the symbolic order at what are represented as its very roots [...] [it] remind[s] the horrified reader of the mother's destructive power" (Miller 92/105–6). Miller summarizes that "to narrate infanticide is to explode traditional conceptions of motherhood and the maternal [...] to explode tradition itself [...] [a] refusal [...] of exile to the margins of language and literature" (Miller 111/125). Masino's unconventional, idiosyncratic, if not straightforwardly negative vision of maternity and renunciation of the maternal – in *Nascita e morte della massaia*, she famously described the process of giving birth as a "dirompersi dell' individuo" – and her avant-garde sensibility and awareness of the role of women and gender/sexual issues,[15] are particularly in tune not only with Re's, but also with Miller's more general critical comments.

14 Mary-Kay Miller, *(Re)productions: Autobiography, Colonialism and Infanticide* (New York: Peter Lang, 2003), 4. It should be clarified that Miller's interpretation of "reproduction" goes well beyond its meaning of "women's reproductive role in society." For example, the author observes that: "In the realm of literature, until relatively recently fiction and poetry alone represented artistic creativity and invention, whereas a genre such as autobiography, considered as an imitation or reproduction of actual events, represented an act of self-indulgence or signaled the onset of decrepitude and has generally been accorded a lesser literary status." (Miller 5/6) In this light, the fact that what Masino writes in the quoted letter to her parents in *Io, Massimo e gli altri* partially reappears in fictional form in *Fame*, could still corroborate, from yet another angle (with her movement from autobiography to fiction), her "infanticidal" "refusal of an exclusive and devalued reproductive role" (Miller 125) and her desire, instead, to produce a "new and autonomous subject and text" (106).

15 Allison Cooper aptly writes about Masino's "innovative presentation of the modernist crisis of consciousness in light of the psychological and socio-political implications of gender" (382).

Taken together, Re's and Miller's observations are useful for an additional interpretation of Masino's short story *Fame*. It is tempting to view the double infanticide narrated in *Fame* not only historically, as a clear attack on or form of symbolical resistance to fascism's domestic ideology and alimentary policy, but also aesthetically, as a figure for the elimination of a "reproductive burden" that may prevent the woman/artist from achieving her autonomous, creative potential; one that shatters the traditional image of motherhood and simultaneously affirms another kind of creator/producer. Instead of lingering passively on "the margins of language and literature" (Miller 125), this new, emancipated woman would instead be able to fashion a productive literary role for herself and finally claim a more active presence in the traditionally male-dominated center.

It is also evident, however, that while this kind of hermeneutical scenario fits Masino's 1939 novel, it may not be immediately relevant to her 1933 short story. In fact, the perpetrator of the infanticide in *Fame* is not the mother who, as the initial dialogue between her children reveals, is already dead at the beginning of the story ("La mamma è morta. – Perché è morta? – Di fame" Masino [41]) but, more problematically, the father. Dealing with male, paternal rather than maternal violence, in fact, may substantially counter Re's insights on abortion and render irrelevant Miller's poignant remarks on fictional infanticide as a form of feminine resistance to, or liberation from, reproductive duties. Finally, it may undermine infanticide's subversive significance in terms of language and gender.

And yet, given Masino's biography and characteristics, it is unlikely that in *Fame* she was exclusively interested in narrowly depicting a "paternitá [...] sofferta" (M.V. Vittori in Masino 1994, 27) without addressing at the same time the parallel implications of the suffering of a starved mother, thus challenging codified gender roles. Does this crucial detail in the plot of *Fame* about the sex of the killer truly invalidate what seems to be an otherwise convincing and satisfactory interpretation? Could it be that the substance of the interpretation suggested above for *Fame* may somehow remain valid and appropriate in these radically modified narrative circumstances? I am inclined to think so.

One may note, for example, that as Masino focuses on the representation of the tragic challenges and decisions that a starved father has to face, she also necessarily forces us to think of the paternal and, as a direct consequence, of the maternal[16] in a radically different way. By choosing to depict a father who kills his children, she not only questions the notion of fatherhood in itself[17] but, by abolishing "the very paternity that patriarchal institutions would represent as having transcendental value and thus as being transcendentally immune to disruption or abrogation,"[18] she reacts against patriarchal authority, be it that of fascism or that of the male-dominated Italian world of letters in the Thirties. At the same time, she subtly promotes, if not the value of maternity *strictu sensu*, at least the production and power of an alternative feminine discourse such as the one that the author herself is "showing off" by writing her shocking and scandalous story.

From another perspective this argument is further strengthened if we observe that, in *Fame*, Masino could also be establishing a dialogue with, and critically reacting to, a European historical, cultural, and literary tradition which identified, with very few exceptions, men, rather than women, as the most frequent authors of "infanticide texts" in which they tended to represent female, rather than male, characters as those associated with the act of infanticide. As Helga Stipa Madland notes, writing about Goethe's *Urfaust* and Schiller's *Kindsmörderin*, "the dominant images of women, the concepts and ideas of what it means to be a 'real' woman were and still are the products of masculine fantasy" and, even more significantly, "infanticide [fiction] [...] still served the same old purpose: that of controlling and subjugating women [...] It is about the intimidation of women [...] in the 18th century [...] [and it] proved to be an excellent and timely vehicle to reassert patriarchal authority."[19]

16 According to Maria Vittoria Vittori, "maternità" is also among the "temi privilegiati [...] dell'opera dell'autrice" (in Masino 1994, 27).

17 Once a man kills his children, he can no longer be defined a father.

18 Samuel A. Kimball, *The Infanticidal Logic of Evolution and Culture* (Newark: University of Delaware Press, 2007), 23.

19 Helga Stipa Madland, "Infanticide as Fiction: Goethe's Urfaust and Schiller's *Kindsmörderin* as Models." *The German Quarterly*, vol. 62, n. 1 (Winter 1989), 27–38, (31, 34–5). These are writers, by the way, with whom Masino was almost

While I am not suggesting that the social and ideological circumstances in eighteenth century *Sturm und Drang* Germany are the same as those in twentieth-century fascist Italy,[20] it seems fair to observe that a span of two hundred years is, after all, not such a substantial period of time, either with regards to women's possibility of accessing and/or transforming cultural codes or, conversely, to men's appropriation of certain themes and motives (i.e., the fictionalization of infanticide) for "controlling and subjugating" purposes (Madland 31). Therefore, in the few pages of *Fame*, Masino may actually manage to achieve a very significant objective: she reverses a well-established literary tradition and actually "transform[s] the cultural code" by subversively assigning a role which historically had been largely reserved for women and mothers to a desperate and disempowered man, one who ends up being a total failure both in his paternal and creative efforts. At the conclusion of *Fame*, the paternal (together with the patriarchal) is fatally weakened and, by "giving birth" to her story while its male protagonist is literally killing its characters, Masino implicitly affirms her own independent, non-subjugated feminine discourse and empowerment, redeeming at the same time a hidden (or fictionally dead) female voice that is, in reality, still desperately hungry to be heard.[21]

---

certainly acquainted. As Vittori notes, as a teenager she had already "divorato Goethe, Shakespeare, Ibsen, Gogol, Cechov, quasi tutta la letteratura romantica." (Masino, *Colloquio* 13).

20 This said, as far as the representation of femininity is concerned, there are some superficial and yet peculiarly impressive similarities between these two periods. Think, for example, of the standard image of woman proposed by fascist ideology (the required passive, modestly learned "angel of the heart," who literally drives crazy Masino's *massaia*) and, at the same time, that "the dominant image of femininity in the latter half of the eighteenth century is 'die Empfindesame,' the beautiful soul who is a refined and decorative domestic appendage, whereas 'weibliche Gelehrsamkeit' [feminine learning or erudition] is ridiculed and scorned" (Madland 30) or, else, at the fact that Goethe's *Urfaust* is pervaded by the ideal of the "angel in the house."

21 Significant, at this regard, also the conclusion of Masino's *Periferia* (1933) in which a girl (Nena) destroys the objects and tools of the (male) workers who are building the new streets and the new neighborhoods: "lei è il sole in guerra contro gli uomini e la loro civiltà." (230)

If *Fame* deals with paternity and the consequences of the absence of real food, it also undoubtedly addresses an even deeper, quintessentially feminine hunger for achieving a creative form of literary maternity, as opposed to a merely biological one. Words such as Ellmann's "the immolation of the body is rewarded by the gifts of words [...] the expression of the word requires the repression of the flesh" seem to provide further corroboration of a fictional situation in which the pre-textual, initial death by starvation of the mother is eventually compensated for by the conclusive emergence of a strong, feminine text and voice, at the expense of a now silent, incapacitated masculine one:

> – Se lei mi fa dare una zuppa calda, dopo le racconto un bel fatto.
> Gli portarono una scodella di brodo: lui tentò inutilmente due o tre volte di ingoiarne un cucchiaio, buttò tutto in terra con ira e si mise a singhiozzare:
> – Fatemi mangiare, fatemi mangiare. Ora come faccio se non so più mangiare! (46).

If, at the beginning of the story, the "tema" ("Il leone") written and read by Mario, Bernardo's son, indicates the possibility that "men can creatively express themselves" – even though, when they do, they ironically reinforce the stereotypical notion of "woman as food" ("vorrei tanto che [il leone] si mangiasse mamma mia quando mi picchia," [42]) – in the end, such a possibility has completely disappeared. The fact that Bernardo no longer knows how to swallow or eat ("non so più mangiare") means also that he can no longer bring the exterior to the interior. He cannot process either food or words as oral objects and, as a final consequence, his own and, arguably, all men's oral dominance, is questioned and contested. The final irony is that while Bernardo, in the end, will never be able to tell the "bel fatto," the female narrative voice has successfully and authoritatively completed her own version of precisely such a fact.

Given that a "successful" double infanticide, perpetrated by a desperate, hungry father who is unable to provide food for his starving children, lies at the narrative core of Masino's *Fame*, an "infanticidal logic"[22] or, perhaps better, a series of symbolic "attempted infanticides" is also the motif around which the whole plot of Bontempelli's *La fame* is constructed. This logic, in fact, is the motivation behind Barbara's return to her homeland (the mysteriously located "isola Savaría") from her South-African exile and her progressive transformation from an apparent peace messenger in the first two acts into a sort of furious, revengeful angel of death at the end of the third. Not only does the audience find out in the first act of the play that she had been abused by Reno when she was only a twelve-year-old orphan in Savaría ("Io ero rimasta a girare per le strade. Un certo giovane ne ha approfittato violentemente: avevo dodici anni." [163]) but also, in the third act, that her father did not particularly love her and left her alone as a child to go in search of her mostly absent mother (Barbara: "Ma tu hai voluto piú bene a lei che a me [...] per questo quando sei andato a cercarla non ti sei accorto che io rimanevo sola. Lo sai, che ho avuto tanto freddo, che ho avuto tanta fame quando tu non c'eri piú?" [187]).[23] As if this were not enough, Barbara was also starved and forced to leave Savaría by the sadism of her old servant, Sira. By refusing to feed her even in times of abundance and starving her when she was a little girl, Sira irrevocably jeopardizes Barbara's childhood, "infanticidally" depriving her of well being as a child (Kimball 19), and shares responsibility both for her departure and escape and, later on, for her return and surprising final act of violence:[24]

22 On the notion of "infanticidal logic" see Kimball: "The infanticidal [...] takes as its subject not the killing of infants and children but the logic, structure, and economy of that which is infanticidal whether or not there is actual death at end." (Kimball 20). The reasons why I claim to individuate this logic in Bontempelli's play should become more evident in the following pages.

23 On the figure of the absent mother and its connection to infanticide, see Miller: "The mother's 'absences' – moments when she becomes foreign, alien, incapable of taking care of her children, of feeding them – underline her non-nurturing character, her 'unsuitability' for motherhood" (Miller 120).

24 Tangentially, Sira's "infanticidal" role in Bontempelli's *La Fame* confirms what we previously observed about a traditionally gendered representation of infanticide (i.e., a male author and a female murderer).

> BARBARA: "Tu [Sira] mangiavi, bevevi, con quello. [...] Io mi sono spinta verso voi, ho detto soltanto: 'anch'io'. Vi siete voltati, lui s'è messo a gridare: 'che ci fa qui la ragazza? Non la voglio, mandala via' [...] E tu lo hai ubbidito subito, mi hai detto in fretta: 'Vai di là, torna di là e dormi'. Io gridavo: 'Ho fame anch'io'. [...] m'hai detto: 'te ne lascerò un pezzo, per più tardi' [...] niente, non c'era più niente, tutto sparito, neppure un pezzo di pane; l'odore della carne che mi dava nausea e faceva la fame più forte" (Bontempelli 194–6).

It seems thus fair to suggest that *La Fame* could be also interpreted as a "drama of infanticide escaped" (Kimball 247), one which apparently dramatizes some of the consequences and effects of such a failed act of violence against the female protagonist of the play.

Secondly, one may observe that Barbara's hunger, first real and then metaphoric, as an appetite for revenge, is paralleled, at least until the play's resolution, by her association with food.[25] She is linked with food at the beginning of the play and, later on, when she offers and sacrifices herself for the sake and the well being of the citizens of Lido and Vetta, the two villages in Savaría that are on the verge of declaring war with each other. Here is Reno, talking to Orazio, the leader of Vetta, about Barbara: "Si ficcava nelle case, portava da mangiare anche a chi non ne aveva bisogno. [...] Per tre giorni ha rimpinzato di pane e di frutta tutte le ragazzine di Lido. [...] i pescivendoli [...] han cominciato a esaltarsi, a seguirla per le strade, a domandare consigli" (Bontempelli 1947, 151–2). For almost three acts, we are led to believe that Barbara represents a peaceful solution to the impending confrontation between the island's two warring factions.

The strong Christological allusion whereby Barbara is a kind of "female Jesus" and a figure of the "host," simultaneously a victim whose flesh has suffered and a woman-savior who, at least for a while, brings the promise of nutrition and speaks the word of peace and forgiveness to Savaría, is hard to miss here. This allusion becomes even clearer in the following confessional

25 Maggie Kilgour notoriously wrote that women "are traditionally associated with food" (Kilgour 243).

monologue in which Barbara, addressing her dead father, in whose death Orazio was apparently involved, ponders the idea of "perdono":[26]

> Certo tu non potevi pensare che io avevo potuto vivere, se sai che avevo sofferto tanto. [...] Vedi, a te non riesco a dire la parola "ti perdono," ma questo non vuol dire. Se tu non fossi morto ora saremmo felici, non ci si penserebbe nemmeno a perdonare o non perdonare, Quando si perdona è per riempire un vuoto. A Reno potevo perdonare. Penso che tu non volevi che perdonassi a Orazio? Che forse è la causa della tua morte. E a tutta quell'altra gente? A tutta l'altra gente, quelli che mi cacciavano come si caccia per divertirsi, non per avere una cosa da mangiare. Ma quando si perdona, sai? Si perdona tutto a tutti e allora anche a Orazio. Se ci mettiamo a pensare non perdoniamo piú nessuno (188).

This passage is intriguing because, besides the reference to an ecumenical act of forgiveness and to the idea that "perdonare" is equivalent to the gesture of "filling a void," as if the act of forgiveness were a form of eating itself, Barbara explicitly refers to herself as a "cosa da mangiare." As one consumes the host in order to find peace and be forgiven his/her own sins, one may similarly consume Barbara's body as a source of spiritual nutrition. Barbara, in turn, repeatedly attempts to persuade everyone and, in particular, her own divided self, that forgiving, like eating, is a fundamental human need and that, as a consequence, by forgiving everyone rather than by taking

26 It is certainly encouraging for this comparative analysis that the motif of "perdono" surfaces also in Masino's *Fame*, just before the father strangles his son ("Sei stato cattivo Mario, ma ti perdono," [45]) and that the exact association between "forgiveness" and "eating" appears in another one of her short stories, *Nozze di sangue* (also written in these years but published in 1947), ("Si mangia forse il perdono? [...] Ma il perdono che cibo è? Non ne avete mai sentito voglia? Neppure immaginate che sapore abbia, e come si compra e a che prezzo." [134]) Barbara's association with a peace-seeking Christological figure and her almost supernatural lightness surfaces elsewhere in the play: Barbara: "Sono le due città dell'isola Savaría, la nostra patria, che è tutta bellissima e sarà sempre piú bella se penseremo a coltivarla con amore [...] con tanto soffrire si diventa leggeri, e si vede tutti gli altri come sono pesanti, e fanno compassione [...] le cose vanno fatte a ognuno per il bene di tutta Savaría. [...]" Bruna [addressing Sira]: "Ma è tanto buona [Barbara] lo sai? Tanto, con tutti. Lo capisci quanto bene ha fatto oggi a tutti? Ha messo la pace. Ha salvato tanti dalla morte. È come un angelo" (Bontempelli 1947, 180/182/184/186).

revenge, she will be able to satisfy her own hunger, or "riempire" her own "vuoto": "Barbara: forse cercare di vendicarmi [...] invece questo è brutto, non va fatto [...] quando uno si vendica [...] siamo di nuovo tanto lontani da Dio [...] ora sanno che Barbara vuole perdonare" (182–3).

At this point, it seems that, as with Masino's story, Bontempelli's play is dealing with the question of the relationship between hunger, infanticide, and representation. *La fame* addresses the connection that links a woman's body to language and expression, the "sacrificial" quality of language itself,[27] and, as I would like to ultimately argue, the fundamental inefficacy, powerlessness, and subordination that such a feminine discourse continues to present. This possibility may be further strengthened if one recalls, once again, Maud Ellmann's observation that "the expression of the word requires the repression of the flesh" (Ellmann 47) as well as Simone Weil's remark that "to starve is to renounce the past [...] because it is to rid the body of its larded history" (qtd. in Ellmann 88).

The perennially hungry Barbara, by trying to forgive and, therefore, to forget and put behind her the burden of her painful past, finds herself undoubtedly in a "starving" condition. At the same time, however, the gesture of forgiving (and, thus, forgetting) is paradoxically that which could satisfy her hunger because, in that way, she would be able to "fill her void." But if Barbara is not able to forgive and forget her past, she cannot satisfy her hunger and is still condemned to starve. When read against the background of Ellmann's and Weil's observations, the implications of Barbara's starving situation take a peculiar twist in Bontempelli's play. If Barbara's starvation is a major ingredient in the the plot, and her history of hunger and sacrifice temporarily provides her discourse with power, weight, and

27 Ellmann observes that "language is a sacrificial order, exacting every pound of flesh, and [...] it is woman who is doomed to make this 'final payment.'" (Ellmann 53). In addition, Kimball's remark that "the Eucharist brings the infanticidal violence of the body's hunger as well as of the Spirit's Word to the very tip of the tongue [...] in the Eucharist [...] Jesus transforms pagan sacrifice [...] into an absolute self-sacrifice, a sacrifice of the Logos by the Logos" (Kimball 28), also seems to resonate with the story of Barbara in Bontempelli's play.

authority when she initially arrives at the island,[28] the relevance that her words may have acquired since the moment of her appearance are progressively "erased" by her increasingly erratic behavior, which culminates in her final, homicidal act.

In the end, in fact, the "angelic" (186) and thoughtful protagonist depicted for most of the play leaves the stage for someone who is "in grande esaltazione," (200) a progressively crazier ("è impazzita" 197), hysteric, bestial, and homicidal woman ("come una belva" [197]), someone whose words can no longer possess any kind of authority, power, or efficacy.

In the climactic scene, Barbara reveals her inability to forgive and, disregarding the guiltier and more important male characters she could have chosen as targets of her revenge (Reno, Orazio), she strangles a secondary, female one, Sira, as if she were symbolically directing her rage onto her absent, frivolous mother and reversing the "infanticidal" violence she had to endure as a child. By doing this, she also condemns everyone, including herself, to perpetual hunger and implicitly reasserts "the violence that is inextricable from the [female] body's appetites" (Kimball 237): "non è vero niente, ammazzatevi, odiatevi, non importa piú niente, non conta, non conta fin che c'è nel mondo la fame." (200).

If Barbara had remained prey rather than predator, if she had been able to forgive everyone and maintain the edible and nutritive qualities associated with her host-like role, she might have been able to affirm the power and authority of her word on the island and put an official end to her physical and figurative belonging to a marginal, "colonial," third-world

28 In this sense, Barbara may be added to the list of those literary "heroines who use biological hunger to obtain the 'authority' necessary to write [or tell] their own stories," in Donnalee Frega, *Speaking in Hunger. Gender, Discourse and Consumption in "Clarissa"* (Columbia: University of S. Carolina Press, 1998), 1. The second act ends precisely with several references to the power, relevance, and almost prophetic wisdom of Barbara's words: "Terzo Lidese: Si, dite voi che cosa dobbiamo fare; Reno: Signora, credo che nessuno potrà fare diversamente da quello che voi avete detto; Barbara: Dovete tornare ognuno a fare le cose che ha sempre fatte, il pescatore pescare, il contadino coltivare la terra, e cosí ognuno il suo mestiere [...] E vi dico ancora che il mestiere giusto di Orazio è comandare" (183–4).

exile.[29] Her peace-oriented message would have been affirmed, and the impact of her authoritative feminine voice and her general "savior-like" qualities emphasized. Finally, by forgiving and, sanctioning her sacrifice and renunciation of violence, Barbara would have resolved the conflict and moved one step closer to feeding the various "hungers" that affected both her and the others.

However the story does not play out this way. The rational, conciliatory attitude, and diplomatic, leader-like (one could now add "masculine") qualities that Barbara displays during the first two acts of the play cannot ultimately be sustained. Her most basic appetites and irrational, violent feminine instincts unavoidably emerge, relegating her original, constructive discourse back to the margins. The hunger motif and its implications in *La fame* only apparently challenge the well-established idea of a feminine discourse that, instead, continues to be perceived as marginal, unstable, and ineffective. Granted, the play begins by offering us an idiosyncratic perspective on female identity, one that Bontempelli could have easily derived from Masino's own persona and work through the innovative figure of an apparently strong, "dangerous" and empowered woman coming from the periphery of the world to save her mother country, someone whose words may have a very concrete effect in the microcosm of the island and contest a widely established male oral dominance. But it then ends by radically undermining that same fresh and unusual image of femininity.

As the curtain drops in *La fame*, the audience is confronted with the depiction of a(-nother) failure which mirrors and subverts that of the desperate father/killer in Masino's story: this time, the image is the one of a disempowered, homicidal woman who was not able to adhere to her own words and intentions. The existence of any alternative is questioned: either when woman is food (i.e., when she is prey, being eaten) or when she eats (i.e., when she displays her predator side, her aggressiveness), her words

29 For a while, it seems that Barbara may actually be the bearer of some reverse civilizing mission, one that paradoxically originates from the margins, from a distant, colonial third world (South Africa) and intends to bring peace and progress to a belligerent and litigious mother country. The conclusion of the play implicitly affirms the unlikelihood that such a subversive dynamics may ever take place.

and actions are destined to have little or no impact in terms of changing the *status quo*.

In this sense, it is telling that, at the end of the second act, what I have previously labeled Barbara's "prophetic" voice utters also a peculiar sentence such as: "E vi dico ancora che il mestiere giusto di Orazio è di comandare" (184), when Orazio, let's not forget, is the main suspect in the elimination of her own father! Barbara momentarily "came back" from the margins, from her physical and psychological exile, bringing her apparent wisdom and superior, enlightened vision (Barbara: "No, voi uomini finireste ancora col parlare di cose pesanti" [185]), only to be eventually cut down to size to a stereotypical role for a woman and placed into the even worse existential entrapment of her "follia."

In the end, as far as the play's representation of femininity is concerned, one is left to wonder if Bontempelli's notorious notion of "return to order" could perhaps be both "abused" and extrapolated from its standard context in order to be applied to a situation in which, after the illusion of a temporary threat, the traditional centrality of masculine discourse and patriarchal authority ultimately remains unchallenged.[30] If Masino's alimentary discourse in *Fame* challenges an androcentric mode of representation and the dominance of male expression with the memorable figure of a man being almost simultaneously stripped of both his fatherhood and his ability to eat and utter words, Bontempelli's *La fame*, with a "trademark" dynamic of "return to order," ultimately reverses and ideologically questions Masino's

30 As Allison Cooper reminds us: "For Bontempelli [...] the notion of a return to order was closely tied to what he considered the profoundly incapacitating 'malattie spirituali' of the nineteenth and the early twentieth centuries. Aestheticism, Impressionism, Idealism, Freudian psychology, and the democratic spirit were, in his opinion, different symptoms of a 'paurosa decadenza' that prized appearance over substance, intuition over perception, the unconscious over the conscious and finally, individual or collective interests over greater laws or truths" (Cooper 380). The reaffirmation of traditional gender roles in *La fame*, in turn, cannot but contribute to that "greater cultural stability" which such a "return to order" was supposed to achieve.

position in terms of gender through the figure of a disempowered, perennially hungry woman whose words cannot be taken seriously.

At the conclusion of this analysis, although one cannot affirm with absolute certainty that Bontempelli adopted or was directly influenced by Masino's discourse on hunger and starvation, there should be few doubts that he was quite aware of it. In *La fame*, he emptied those discourses of their subversive potential, exploiting them instead to re-establish that "greater cultural stability" which apparently, for him, could only be achieved through a traditional, male-dominated perspective on language and expression.

3.2 Feeding on Stale Bread and Books: Paola Masino's *Nascita e morte della massaia*

> The progress of an artist is a continual self-sacrifice, a continual extinction of personality.
>
> — T.S. ELIOT, "Tradition and the Individual Talent"

> Pensava e si mangiava le unghie; finite le unghie e i pensieri, masticava tozzi di pane e sfogliava libri in cerca di altro nutrimento.
>
> — P. MASINO, *Nascita e morte della massaia*

Paola Masino's previously mentioned novel *Nascita e morte della massaia* (1939; published in 1945; from now on *Nascita*) depicts the surreal portrait of a woman who – reacting in a paroxysmal way to her mother's exclamation: "Verrà il giorno in cui m'avrai fatta morire di crepacuore" (13) – initially refuses to "come to light" and chooses instead to live the first part of her abject life in a dirty trunk scattered with books and organic/inorganic debris, feeding herself with "groncioli," pieces of stale bread, and, figuratively, with those same books.[31] Emerging from her "baule" (only to please her mother's wish), after a hilarious, implausibly successful but inevitably doomed-to-fail attempt at behaving as a perfect bourgeois, married housewife, she goes through a process of ascetic refusal and renunciation, ("si spogliò sempre più dei suoi beni materiali e delle sue possibilità spirituali per distribuirle a chi ne aveva più bisogno di lei" [242]). In time, she expresses a will to "smateriarsi" and "annullarsi" ("Si sentiva sempre più distratta e leggiera e senza più attesa" [270]), until the day of her long-wished-for, actual death (returning to her original trunk-womb-sarcophagus).[32]

Building on the contributions of gender studies and feminist scholars, I would like to suggest in the following pages a reading of *Nascita* which emphasize the traits and concerns that the text may share with a more encompassing and non-compartmentalized (i.e., not exclusively gender-restricted) modernist aesthetic. This, perhaps, may appear a counterproductive or even impossible task since we are dealing with a novel which deals mainly, and in a "mock-heroic" register, with the stifling condition of women during fascism (the anonymous *massaia* of the title, as a sort of "Everywoman"), their progressive deprivation of freedom, sexuality, and creative impulses, and final reduction to a "perfetta macchina domestica" whose only liberation is death.

31 In a situation which may partially recall the one of Kafka's Gregor Samsa, the *massaia*, by eating "groncioli," equally "refuses to eat the food which establishes and maintains the family as a single body." Since, like Gregor, she eats "what the family has rejected as inedible," "[s]he becomes inedible to the family, an unclean and monstruos non-form ..." (Anderson 31–2): "Dopo anni di tale martirio la sua vita divenne ancora piú estranea ai familiari [...] la famiglia non badava piú a lei che come a un mobile" (16). Anderson's observation that "in attempting to deny the body, the modernist produces an antibody which withdraws from the traditional arena of male privilege, authority, and responsibility" (37) moreover, seems to aptly fit also to the general context of *Nascita*.

32 Maud Ellmann, dealing with Richardson's *Clarissa*, notes that "writing is itself a process of sarcophagy, because Clarissa's epistles consume the very body that indites them" (83). Similarly, while the *massaia*'s body is "eaten" inside the trunk, her own story proliferates.

And yet the plot of *Nascita* and the presence in the text of certain motifs and questions which I have previously associated with a quintessential modernist rhetoric indicate that this approach may not be completely out of place. Think, for example, of the issue of "self-sacrifice" and "continual extinction of personality" mentioned by Eliot; of the specific trope of the book as a kind of food (and, conversely, of food as a kind of language) that emerges in the first page of *Nascita* and, more generally, of the sub-theme regarding the semantic field of productivity (or lack of it), be it food, babies for the *patria* or, self-reflexively, language itself. Think also of the "autobiographical obsession" that characterizes *Nascita*,[33] at the *massaia*'s own progressive, ascetic gestures of renunciation, death-wishes and silences ("morí quella stessa notte, senza aver piú aperto bocca o chiuso gli occhi" [277]) and of a, by now familiar, pattern of self-consumption that absorbs "the writer into [the] text" and simultaneously gives birth to a text that has consumed him (or her) (Heywood 77). Consider then, in the same light, the already emphasized connection between similar forms of corporeal self-effacement and the notion of a modernist "language crisis" (Anderson 29). And, finally, add to this that the allusive, esoteric and allegorical prose that characterizes so many modernist texts seems privileged also by Masino ("*Nascita* riguarda le casalinghe quanto *Moby Dick* gli studiosi di balene").[34]

As discussed in the previous chapters, issues similar to those characterizing Masino's aesthetics and poetics not only have haunted, in their own peculiar way, the artistic vision of key modernist figures such as Eliot and Kafka but, to remain in Italy, besides the frequently mentioned Bontempelli and Pirandello, even the vision of Palazzeschi and the futurist avant-garde.[35] Those situations in Palazzeschi's *Il codice di Perelà* and

33 Writing about *Nascita*, M. Mascia Galateria notes that "Un'ossessione profondamente autobiografica e vera fin nei minimi particolari, diventa la storia di un fallimento." (Mascia Galateria 3).

34 Silvia Giacomoni, Introduzione a *Nascita e morte della massaia* (Milano: La Tartaruga, 1982), 5–10, (10).

35 Masino's (and Bontempelli's) indebtment to the more irrational, esoteric sides of the futurist avant-garde is worth further exploration. Masino's links to the the futurist

: riflessi or in Marinetti's *Le Roi Bombance* that allude to various instances of "enclosures" and subsequent, enigmatic corporeal obliterations in relation to questions of aesthetics and artistic creativity may come easily to mind at this point.[36] At the same time, it seems productive, when dealing with a novel such as *Nascita* and its "imprisoned" protagonist, to think again of the equally surreal situation of the voluntarily caged, self-starving protagonist in "A Hunger Artist" who, as Ellmann observes, "has to starve in order to perfect the work of art" (Ellmann 59).

If, as Rorandelli noted, the *massaia*'s behavior intends to criticize fascism's notorious promotion of women as "reproductive machines" geared to provide children for the "Mother" country and, we should add, to undermine and dissent from its autarchic, bourgeois food politics, it may also allude to the protagonist's (and, by extension, her author's) own effort and struggle to deal with "creation" in a wider sense.

It is, therefore, the recurrence of what should by now be a familiar pattern of "literary food aversion," that I intend to trace and discuss in its particular manifestations in Masino's novel. By focusing on the widespread food-imagery in the text, on the progressive "extinction of personality" which marks the life of the *massaia*, and on the contradictory opposition between notions of productivity and sterility, I intend to suggest that *Nascita* too, perhaps in spite of some of its most apparent textual evidences, may also be read as an allegory for the difficulties associated with literary

---

avant-garde, after all, were already noted by C.E. Gadda in his biting review of *Monte Ignoso* (1931), when he wrote that Masino's style is "ai limiti di un futurismo deteriore, tutto notazione dell'immediato percepire e niente espressione del profondo apprendere" (in Airoldi Namer 2000, 185). Additionally – and although the contexts are different – it is at least noteworthy to quickly remember that 1. If "Da bambina la massaia era polverosa e sonnolenta" (13), a "casalinga polverosa" is also the protagonist of Rosa Rosà's *Una donna con tre anime* (1917); 2. that a comparison between the *massaia* and "l'automa-femmina" della marinettiana *Elettricità sessuale* has already been suggested by Airoldi Namer (Airoldi Namer 2000, 166) and, finally, 3. that issues of sterility/creativity are crucial in Bontempelli's *Gente nel tempo* (1936) (the character of Nora is sterile).

36 For a detailed analysis of the notion of *emboîtement* in Masino, see Airoldi Namer's article.

creation and, ultimately, follows the sort of consumptive "logic" I have individuated in other modernist texts.

Masino's last novel is often interpreted by critics as a "storia di un fallimento senza ritorno" (Mascia Galateria, 6), correlating to the limitations of the feminine condition under fascism; as the exploration of an impasse, of a double impossibility consisting of the urge to rebel and the pressure to conform (Re 1995, 92–4); as the representation of the tragic, reductive loss of "identità individuale a favore della propria funzione familiare e sociale."[37] In short, many scholars agree that the novel may be regarded as an allegory in part protesting fascism censorship and suppression of women's creative powers, a sort of literal gravestone and defeat of the possibility of feminine imagination and creativity. Lucia Re's affirmation: "La Masino dà quindi un impronta specificatamente politica e di genere al tema dell' 'artista alla fame' (o della fame) ereditato da Kafka. La Massaia non solo non trova il nutrimento spirituale di cui l'artista ha bisogno, ma, dovendo materialmente dare nutrimento, l'essere artista le è precluso" (Re 2005, 180n), summarizes this kind of approach and reading.

These interpretations, as they appropriately place *Nascita* in its historic-cultural context, undoubtedly capture the fundamental components and the spirit of the text. However, just as the novel subtly and ironically questions "il discorso fascista sulle donne" (Rorandelli 93n) and explores the dramatically reductive and weak role of women during the decades of the regime, it effects an additional and successful (counter-) discourse, produced by a woman in those same years. In my estimate, an additional, perhaps less dramatic and tragic message may emerge from this depiction of a "failure," a "preclusion" and a "defeat": a message that alludes to a fundamental faith in the resilience of the powers of imagination and artistic creativity that may be traced in the novel.

While exploring this possibility, it may be useful to look at the deeper implications of notions of "failure" and "defeat." That is, I would like to consider that, at least at a tropological level, this failure, this literal "dying

37 Tristana Rorandelli, "*Nascita e morte della massaia* di Paola Masino e la questione del corpo materno nel fascismo," *Forum Italicum* 37, 1 (Spring 2003), 70–102, (85).

through words" (Heywood 83) contributes to both to the final authority of the *massaia*'s discourse and, more tangentially, to a sort of figurative rebirth; to a fundamental confidence in the lasting and empowering capacities of language and fiction ("fino alla nascita nuova" [76]).[38]

As the title itself indicates, metaphorical notions of birth and creativity as opposed to those of death and sterility provide the backbone for Masino's novel. The fact that the *massaia*'s mother, for example, when her daughter eventually dies back in her trunk, "urlava come quand'ella era nata" seems to confirm and intensify, at the end of the novel, the confusion and the overlapping of two events (death/birth) and two ideas (self-cancellation/procreation) that are antithetical but that, nonetheless, are linked through the narration.

A couple of similarly "antithetical" considerations should be made before approaching the text more closely. The first is that "the writing of a book has always traditionally been compared to the birth process" and that "for psychoanalysis the book is a maternal symbol," one that – as Georges Bataille put it – is "the equivalent of the womb."[39] The second is, besides her controversial relation with food, that the most powerful "starving" and truly "anorexic" attitude at the core of the *massaia*'s behavior derives from her firm refusal to procreate, her extremely negative vision of motherhood, which she perceives as an annihilation and disintegration of the self, a literal "dirompersi dell'individuo" (233) and, eventually, the discovery of her sterility.[40] Investigating and reconciling at both the literal and figurative

38 The idea that a pattern of suffering and renunciation is related to a (feminine) discursive authority is discussed in Donnalee Frega's *Speaking in Hunger, Gender, Discourse and Consumption in "Clarissa"* (Columbia: University of South Carolina Press, 1998). As a consequence, it does not really seem that an emphasis on motifs of corporeal sacrifice, mutilation and food refusal in *Nascita* is in contradiction with the logic of a narrative text that someone may call "femminista." What I am suggesting, is that it may be precisely the redemptive quality of such an emphasis that finally allows the female voice to be regenerated and thus even more powerfully heard.

39 Qtd. in Maryline Lukacher, *Maternal Fictions. Stendhal, Sand, Rachilde, and Bataille* (Durham and London: Duke University Press, 1994), 181–2.

40 The close relationship between the refusal of fertility and anorexia is well known from both a scientific and literary perspective. Besides Susan Skubal's assertion that

levels the contradictory co-existence and tension in the text between, on the one hand, the impulse to "give birth and creating" and, on the other, the apparent negation of that same impulse, is the principal objective of the rest of this section.

"Le parole hanno un valore e voi vi ostinate a non riconoscerglielo" (28), the *massaia* tells her simple-minded mother at the beginning of the novel, a few moments before getting out of her trunk. Later on, we shall significantly find her first "avverti[re] il proprio annullamento, che è il modo piú certo di affermarsi" (66) and then asking herself: "che cosa è questo morire se non tornare a un solo grembo che tutto contiene e tutto esprime, il morto e il vivo a un tempo, il nascituro e il trapassato? E il suo esprimersi che sarà dunque? se non una cosa che non abbia comunque materia? un fiato? un suono? una parola, il verbo?" (76).

It seems, indeed, of crucial importance for readers of Masino's novel to understand the meaning implicit in "to affirm oneself through one's cancellation." In addition, it is important to reflect on the protagonist's ruminations about death, birth, life, and, ultimately, the possibility of expression and representation. One could begin by simply observing, for example, that while the *massaia* rejects the idea of procreation, she also wishes to return to the womb-like trunk, where she existed in the first place, and, thus, if Bataille is right in his book-womb association, to "get back to her book(s)," to the dreaming and the peculiar artistic nutrition they provided her.

It should be clear, after all, that the novel, in spite of its protagonist's refusal of "normal" food, motherhood, and her embrace of her professed sterility, manages to show off an (over)abundant, inventive variety of linguistic and narrative devices (or "ingredients"). Sections characterized by third-person narration, theatrical dialogue, first-person diary entries,

---

"the anorexic can't be mother," (74) see for instance Inge Lanslots similar remark that "l'altra conseguenza dell'anoressia è il rifiuto della fertilità e consecutivamente della maternità," in Bart Van den Bossche, M. Bastiaensen, C.S. Lonergan, eds, *Soavi sapori della cultura italiana. Atti del XIII Congresso dell' A.I.P.I.. Verona/Soave 27–29 agosto 1998* (Firenze: Cesati, 1999), 377.

"interpolated narratives" (Re 1995, 92) alternate one with another and, in the end, produce and successfully "give birth" to the text itself.

If, moreover, the reader recalls the autobiographical nature of the text and that, after its composition Masino renounced writing, retired into her private life (her own "trunk"?), and practically gave up all her artistic ambitions ("[after *Nascita*] ho scritto ancora qualche poesia carica di morte. Quanto al resto, ho avuto troppo da fare: ho dovuto vivere e lavorare."),[41] Masino's last novel could be interpreted as a sort of "last supper," eerily similar in its consequences and lasting effects to those following the final meal the *massaia* cooks and supervises near the conclusion of her existential parable.

Under this light, which sees the *massaia* as a double for the figure of the modernist artist and her creative agony, the novel may resemble the final attempt at coordinating and displaying a literary (as parallel to a culinary) "production." It is a sort of "creative gorging" for the protagonist and her author before the actual withdrawing and "starving" (physical and metaphorical) that will follow both the preparation of the fictional meal (which, appropriately, is defined "una disfatta della Massaia" [218], after which her "volontà caparbie di annullamento" [242] are intensified) and the real completion of the writing. As I am trying to argue, however, it is also such a "disfatta" that may simultaneously provide the *massaia* with a more authoritative and convincing voice and constitute the premise for the birth of her literary (and other meta-literary) discourse(s).

A closer look at the *massaia*'s particular relation with food and at the multifaceted alimentary discourse that pervades the text is useful.[42] As the

41 http://www.letteraturaalfemminile.it/paola_masino.htm [Last accessed June 2011].

42 Bart Van den Bossche reminds us that "Il fatto che un personaggio mangi una determinata cosa è altamente emblematico per il suo rapporto con il mondo; gli conferisce una precisa collocazione spaziotemporale, rivela la sua appartenenza sociale, è significativa per la sua posizione idelologica, e offre indizi sul suo stato fisiologico e psichico." (488). Additionally, from a more specific gendered perspective, M.G. Muzzarelli and L. Re observe that "nel rapporto della donna con se stessa, nella defininzione della sua identità e del suo potere e nella sua capacità di espressione estetica, il cibo risulta avere un ruolo fondamentale e fondante." (14).

quotation at the beginning of this part indicates, among the first pieces of information the reader receives about the protagonist are that the act of thinking and the one of an autophagic self-consumption (or self-mutilation) are intimately linked ("Pensava e si mangiava le unghie" [13]);[43] second, that her nutrition is both physical and intellectual ("masticava tozzi di pane e sfogliava libri in cerca di altro nutrimento" [14]) and, third, that, given the kind of food she eats, she obviously places more importance on her intellectual feeding than her physiological one. A strong connection between the alimentary ingestion and the literary/artistic production is thus established in the opening section of the novel.

The next, food-related element that stands out in these initial pages is that the flesh of the *massaia*'s body is compared to the edible meat one can find at the market or at the butcher: "il suo corpo era carne come quella esposta sui banchi dei mercati o appesa nei negozi dei macellai" (14). The perception the protagonist has that she herself is some sort of food to be consumed is subtly re-confirmed when we discover that her family, for Easter, pushes her "come un mobile [...] nella stanza da pranzo tra le torte pasquali e i vassoi di uova sode, a che il prete la benedicesse, povera creatura" (16).

If it is true that in the former passage she is specifically compared and perceived by her family as a piece of furniture and not, explicitly, as a piece of food, it is also true that the absurdly ridiculous process the young *massaia* needs to undergo before entering the dining-room ("la lavavano con la soda, le davano cera sui capelli, petrolio alle giunture, guardavano che la pelle del volto e delle mani non fosse tarlata, le accomodavano una ghirlanda di violacciocche sul capo e intorno al collo e ai polsi gale di carta velina azzurra o rosa" [16]) is strongly reminiscent of the preparation of a specially decorated dish and of the gestures needed to execute a particular recipe for a festive day. Her anatomically detailed, well-washed, oiled, checked-for-faults and garnished body may more closely resemble a

43 On this, see also Anderson's remark: "When the self can no longer ingest and digest the world as food, can no longer turn the raw matter of sensation into abstract concepts, judgments, and generalizations, the subject is thrown back on itself for nourishment, becoming both brute matter and pure spirit." (31).

"sacrificial lamb" (an Easter roast ready for the oven), than an inanimate and limbless "mobile."

This resemblance will be reinforced a few passages later, where, once she is "done," she is placed in the dining-room, among other prepared dishes and alimentary items ("torte pasquali [...] uova sode"), waiting for the traditional food blessing from a priest but, at the conclusion of this section, the image of a real lamb slaughtered for food makes its official appearance in the narration ("Come da un agnello sgozzato si ricava il cibo, cosí voleva che subito dal tramonto nascesse una cosa utile" [19]). The text is telling us that the *massaia* is precisely that lamb and, as such, she represents a symbolic, sacrificial food consumed in order to achieve a still-vague higher purpose and objective. The prophetic allusion to a *massaia* who is also a *messia* is hard to miss here. In both instances, self-sacrifice, death, and a form of resurrection will be essential steps in the creative process and necessary to achieve a most effective spreading of the "word."

Other references dealing with images of metaphorical nutrition reinforce this sacrificial motif in the initial chapter of the novel. A remark such as "Ci hanno chiamato chi sa da dove su questo pianeta e ora dobbiamo alimentarlo" re-proposes more clearly the idea that human beings in general and the *massaia*, in particular, are here specifically to sustain, and therefore, "feed" the planet.

When the narrator informs us that:

> Al mercato cominciò ad amare il cibo perché era un modo nuovo che le si presentava di dare e prendere la morte. Guardava i ventri concavi dei buoi appesi con uncini di ferro alle travi delle macellerie. Dondolavano piano, privi dei loro organi; questi pendevano lí vicino non piú legati al loro alveo naturale ma a strane radici di metallo [...] La bambina traeva la conclusione che anche lei nel suo interno doveva avere qualche cosa di cui il mondo aveva bisogno ... (17–18)

the *massaia*'s topsy-turvy, paradoxical understanding of food as a "death-provider" emerges and clashes with its standard, life-sustaining and nutritive quality.

By associating the hanging oxen and their open wombs and mutilated organs with her own body, she not only implies that her flesh may have a similar feeding function, establishing an analogy between food and her own self and anticipating her negative vision of maternity as a "dirompersi dell' individuo," but also that part of what the world needs will be provided by her eventual sacrifice and consumption.

As a consequence, she describes as "terribile" both the feeling of immortality she has on the occasion of her visits to the butcher and, similarly, her sensation of "non arrivare mai, per quanto faccia, a liberarsi in modo definitivo del corpo che le hanno messo addosso." As the meat of the dead oxen will be consumed and eliminated in order to fulfill its vital function, so the *massaia*'s own body will eventually be disintegrated and be reduced to dust in order to achieve some purpose. What may be equally useful to notice, at this point (beyond the usual, implicit sarcasm of the narrative voice), is not only that this Deleuzian representation of a "body without organs" "incarnates" precisely that condition of dispossession of the bodily self that feminist scholars have linked to women's own historical condition, but also that, in turn, it is reminiscent of the importance Bataille (writing about the ear of Van Gogh), assigns to "self-mutilation as a sacrifice analogous to, and symbolic of, the artist's activity."[44]

For Bataille, as Rosalind Krauss puts it, "the beginning of art [coincides with] the representation of sacrifice, the symbolic correlative of the mutilation of the human body."[45] He continues (with a most appropriate culinary image) by also observing that sacrifice "eats up what could have remained in the chain of useful works" and that it:

> restores to the sacred world that which servile usage has degraded, profaned. Servile usage has made a *thing*, and *object*, of that which profoundly partakes of the *subject*, participates intimately in its nature [...] In the world of things, man himself becomes [...] one of these things. From this fall, in all ages man seeks to escape. In his unfamiliar myths and cruel rites man is, from the first, in search of a lost intimacy (qtd. in Michaud 42).

44 Qtd. in Eric Michaud, "The Real as a Primal Scene (A Note on Art and Devouring)," in *Utopia Post-Utopia. Configurations of nature and culture in recent sculpture and photography, January 29–March 27, 1988* (Boston: M.I.T. Press, 1988), 35–51, (42).

45 Rosalind Krauss, *The Originality of the Avant-Garde and Other Modernist Myths* (Cambridge: M.I.T. Press, 1985), 82. She adds that "Painting is born with man's refusal to reproduce himself, and out of an act of self-mutilation" (83).

I certainly do not wish to suggest that the same level of continuity and complexity that the French philosopher establishes between self-mutilation and Van Gogh's art exists also with reference to the situation of the protagonist of Masino's novel. Nonetheless, at the same time, (adapting Bataille's insights to our circumstances) it is tempting to observe that the *massaia*'s "sacrificial journey" from trunk to grave, from enclosure to a sort of paradoxical freedom, could be read as an attempt to escape a process of objectification and commoditization, which reduces everyone and everything to their utility and tries to re-establish instead a condition of original and creative "intimacy."

In effect, through the *massaia*, Masino seems inclined to wish to break a condition which, in turn, by interrupting the chain of "useful" work as represented, for example, by cooking and eating, is also the premise for any artistic accomplishment. The autophagic gesture of the *massaia*, the perception others have of her as a "mobile" and her own self-perception as a sort of sacrificial food (in both instances, always a product or a "thing"), the refusal to see herself as a complete body, not to mention her ideas on notions of usefulness and uselessness, may indeed reinforce the impression that Bataille's observations on the relationship between sacrifice and artistic creation are indeed relevant to Masino's novel.

The *massaia*'s position "contro l'inutile," by the way, seems particularly revealing if read in this Bataillean context. She is not against the concept of "uselessness" in absolute terms. She is not in a negative "stato di asprezza" towards such a notion in the way the capitalist world of bourgeois production would be. The fact that she is "sempre tesa a cercare il profitto delle cose che gli altri spregiavano" (19) indicates, on the contrary, that her effort (similar to the one of a stereotypical artist), is aimed at finding some positive value (a profit) in what the majority of people despise; that is, in that which others consider useless, a waste of time and not a part of the chain of production.[46]

46 To those who may question my use of a misogynous and male modernist paradigm such as Bataille in order to underline the importance of Masino as a modernist, female writer, I respectfully ask: "Why not?" Isn't my final argument made stronger

The fact that at this initial, still "intimate" stage, old books and "groncioli" constitute her literal and symbolic food is perhaps less of a surprise now. She "eats" them in order to draw some kind of profit out of them, and this process is something the people who surround her are unable to do. If the trunk is a womb-like extension of her mother, books – as Bataille also suggested – are an essential component of this environment in which she can dream and imagine. They constitute a sort of umbilical cord that is necessary for such activities ("leggeva fino all'alba" [16]). When she finally agrees to get out of the trunk, such a nutritive lifeline is abruptly cut off and, as she eventually states, she feels "imbalsamata, dentro" (264).

From now on, books and artistic works in general (by Strawinsky, Goethe, Shakespeare, Leopardi, Rousseau, Tolstoi, Cervantes and Beethoven) are tragicomically able only to provide her with hints on home economics rather than truly inspire her and feed her imagination. Her exit from the trunk marks the end of her dreaming and, significantly, coincides with her progressive exposure to normal food such as the "piatti di spaghetti e larghe pizze alla napoletana" (49) served at the party organized to celebrate her "mock-resurrection." In like manner, she is exposed to the clichéd gestures and stereotyped discourses commonly generated during family meals (72) and to the elaborate culinary skills a truly competent fascist housewife needs to master ("Le insegnerò io un modo eccellente di preparare il merluzzo; suo marito ne diventerá ghiotto" [73]).

All these new foods and alimentary information provide no nourishment for the *massaia*, who does not assimilate them either physiologically or intellectually. The preparation of food becomes a necessary duty associated with "lavori forzati" (84) and cooking is an unavoidable task ("la cucina [...] è una necessità non un evento; un evento, per quanto storico,

---

and even more convincing if it is supported not only by traditional feminist scholarship but also (and especially) by a theoretical framework provided by two eminent representatives of the "other side"? I would imagine that even the most relentless feminist critic should be actually pleased if some of the ideas produced by such a paradigm are appropriated and "exploited" with the specific goal of further valorizing and appreciating the quality of a female writer.

può evitarsi, la cucina non si può evitare" [101]). In this scenario, it is not surprising that a question such as "Di che si nutriva?" aimed at discovering the *massaia*'s new diet as a newly wed in her "giornate deserte e ferme" (63), is destined to remain without an answer and that, as the day of the final meal she has to organize approaches, we actually find out that "Da sei giorni [...] fermenta e crea, non dorme, mangia e dimagra" (95). The appropriation of the requested culinary expertise corresponds to both her progressive consumption and to the withering away of any form of creativity that is not associated with a sterile food preparation: "Dove allora prenderanno posto i pensieri fecondi? Dove, oggi che possiedo divani di piuma e fiori chiusi nelle serre, dove faró io sedere gli elfi che da piccina mi sorridevano tra le muffe del corpo?" (78).

Her previous diet based on stale bread and books corresponded to an abject state of deformity but, nonetheless, it formed and organically stimulated her dreams and creativity. Normal food, on the other hand, is a sign of a forced conformism to a certain, pre-defined domestic lifestyle and rather than sustaining her, it will eventually be conducive to her death wishes and, thus, to her return into the "baule."

The fantastic episode in which the *massaia* meets her younger alter ego and offers her a selection of dishes in her upscale dining-room seems not only to reinforce the relationship between the alimentary and the literary discourse, but also illustrates how both the older and the younger versions of the *massaia* lose a crucial part of themselves (i.e., their irrational, creative side) once they are conquered by the over-abundant signs of bourgeois food and gastronomy:

> Due camerieri si movevano rapidi e offrivano vassoi pieni di odorose materie colorate che la Massaia chiamava maccheroni, fagiano, fagiolini, purea, panna, crema, pesche, moscato e canditi, ma che la ragazza rifiutò tutte con tenacia e impaccio, diventando sempre piú pallida per fame [...]
>
> "Non ti lascio muovere da tavola se non hai mangiato. Se no chiamerò il marito che sa ogni trucco per condurre chiunque alla normalità."
>
> A questa minaccia la ragazza disse: "Groncioli" [...]
>
> "Portate pezzi di pane secco" urlò la Massaia esasperata ai camerieri. "Molto pane secco. Tutto il pane secco che c'è in casa."

> Vi fu nella stanza una pausa piena di sgomento. Poi il maggiordomo fece un passo avanti: "La Signora mi scusi, non c'è pane secco in casa. [...] La distribuzione del pane è regolata in modo che non ne avanzi, che non ci sia inutile spreco [...] Il pane secco non si può fare, Signora. [...] Il pane secco, se la Signorina permette, non è che vecchio romanticismo."
> "Allora" disse la ragazza "morirò di fame."
> "Anche questo è romanticismo."
> "Però" disse la ragazza "mi fa piacere e mi nutre e i vostri cibi giovani fanno un gran chiasso di colori senza nessuna sostanza" (173–5).

"Pane secco," the only kind of food that pleases the "ragazza" and that she wishes to eat, is labeled "vecchio romanticismo" – (as is, by the way, the gesture of "starving to death") – and, as such, does not belong to the category of "cibo razionale e sostanzioso" (174).

Similar to the "ism" and the artistic ideal with which it is derogatorily compared, "pane secco" is a sort of "irrational" substance, one that requires an understanding of time different from the one the "maggiordomo" displays, that cannot be made following a sudden request (unless it is "di cartone dipinto," a cardboard fake) and, finally, that cannot exist in an utilitarian economy of bourgeois production where everything is perfectly scheduled, regulated, and measured in order to avoid waste. Like a true, creative work of art, "pane secco" represents a sort of "useless waste" in the chain of production and consumption, and yet its particular nutritive qualities and its appeal for the "ragazza" stem precisely out of this alternative status.[47]

At the same time, eating "groncioli" and starving are not easy when everyone else around is having a feast (or, extending the allegory, pursuing one's own serious artistic vision rather than compromising with the fashion of the times). Such gestures require persistence and sacrifice. As the *massaia* will later tell her younger double: "anche tu mangeresti il loro cibo, perché morire di fame, credilo, è un'impresa tanto lunga che prima di averla compiuta hai troppe volte il modo di pentirtene" (200). The difficulty of

47 It is tempting to establish an ideal connection between the massaia's "pane secco" and Valentino's "host" in Palazzeschi's *: riflessi*. Both foods, in fact, possess an enlightening and salvifical quality.

this undertaking is the reason why the "signorina" temporarily gives up her ascetic disposition and ends up eating those "cibi giovani" which – like a cheap painting – display "noisy," clashing, flamboyant colors and yet have "no substance:"[48]

> La sera stessa, a tavola fra moglie e marito, la signorina mangiò di gusto la crema di asparagi e chiese due volte il fagiano, ma c'era l'aragosta e fu contenta di impararne il sapore. Si riempí inoltre le tasche di grissini per aver di che masticare fino al giorno dopo (176).

The symbolic price for indulging in such an "abbuffata" and acting only according to rational and utilitarian directions (as her saving of "grissini" suggests), however, is to become too full and, thus – to invoke the painting allusion once more – to lose the promise and potential a white, empty canvas may offer. Unlike "groncioli," the socio-culturally over-abundant, elaborated products of bourgeois cuisine "fill the void" but, at the same time, paradoxically eliminate one's essential "hunger" (as a metaphor for "desire") for dreaming or creating in personal and original ways.

The narrative takes a crucial turn as soon as the final products of such a cuisine have been prepared and the guests are almost ready to sit at the table for their meal. At this point the *massaia* performs the symbolic gesture of taking her elegant clothes off ("questi vestiti non sono più adatti per me" [218]) and putting on a simple cloth ("sacco"), thus literalizing her "disfatta" and (revealing her insanity) simultaneously leaving behind the rational, bourgeois persona she has unwillingly accepted to wear so far. If her new, submissive attire points to the "conformità raggiunta [...] all'ideale fascista della donna" (Rorandelli 83), it is also the appropriate accessory for her new (or, better, rediscovered) ascetic and sacrificial attitude and

48 The "noise" produced by this kind of food and cuisine is such that it prevents the possibility to think: "Cuoco, dì ai tuoi sguatteri di non spezzare le foglie dei carciofi a quest'ora; io voglio pensare. Puliteli a un'altra ora! [...] Tutta la casa ne risuona e io ho l'emicrania!" (206)

merely the most evident, exterior sign of an even more substantial, interior "lightening."[49]

Her stripping gesture initiates the last phase of her "starving enterprise," culminating in a death whose real cause remains vague but that, nonetheless, fits into the anorexic pattern of the novel. In addition, the *massaia* significantly dies "senza aver più aperto bocca," in silence and, allegedly, without having ingested any more food but, perhaps even more revealingly, displaying an "estremo languore" (277), the characteristic sign of a real and metaphoric hunger.[50]

This progressive pattern of sacrifice and corporeal self-obliteration ends here but, unsurprisingly, this is not the end of the story of the *massaia*. Her eminently anorexic, modernist impulse towards self-consumption is necessary not only to define her identity[51] and sanction the authority of her discourse, but it is also a precondition for inspiring new discourses. The "Epilogo" of the novel, with its persistent allusions to the supernatural, despite its continuous ironic and mocking tone, ties together the transcendence of the creativity and the regeneration of the *messia* with the figure of the *massaia* by depicting a parallel resurrection scene and, as such, may reinforce this alternative line of interpretation:

> Molti anni sono passati, ma la fama della dama previdente e benefica persiste tuttora [...] quello speciale becchino [...] raccontava di aver spesso veduta la Massaia uscire dalla sua cappella e, preparata con un po' d'acqua e terra e polvere d'ossa una poltiglia fosforescente, spalmarne le borchie, le lettere, le maniglie e quanti finimenti d'ottone si trovano nel mausoleo [...] A volte si mette a parlare con qualche vicina che abbia anche lei finito di rassettare la tomba al marito (281).

49 Writing about Richardson's *Clarissa*, Frega appropriately notes that "Saintly vitae commonly stress denuding activities [...] As Clarissa slowly turns into a pallid, skeletal saint, she sells her clothes, insisting that they not be purchased by her friends" (106).

50 Lucia Re notes that "[Like Pinocchio] La Massaia ha una dieta [...] scarna e limitatissima. Non prende cibo per non diventare, a sua volta, cibo." (Re 2005, 175).

51 In fact, sacrifice and abstinence, which basically define the *massaia*'s identity, also contribute to "erase" her.

Together with the amused denunciation of the professional housewife curse that affects the *massaia* even after her death, funeral, and burial, one of the first things we learn from this passage is that her fame survives her death. If it is permissible to momentarily extricate the "Epilogo" from the socio-cultural context and to read beyond the ironic intent, these words serve to indicate once again that only after her finite, physical body has been literally and literarily effaced is her infinite, aesthetic re-birth via a fundamental linguistic operation possible.[52] Her own voice will remain present and heard after her death, so that even a humble undertaker (the "becchino") may be moved and inspired to produce and tell a gothic tale about her wanderings in the cemetery. In this particular sense, the death of the *massaia* coincides with a verbal creation and manages to revitalize precisely those imaginative and artistic powers that were basically declared dead at the moment the protagonist abandoned her original "intimate" enclosure.

At the same moment the author concludes her book and her narration, we are reminded by the protagonist's "sterility" and by the undertaker's storytelling abilities that creation is "a model that corresponds to the masculine other" (Lukacher 119) and that "art is only possible through struggle against the body and a world gendered as feminine" (Heywood 77). The imaginary vision of an older Masino, who renounced her creative side and her own work in order to take care of both Bontempelli's body and literary *corpus* may well reinforce such a reading.

It also seems legitimate to add that, at the conclusion of the novel, the *massaia*'s "body disappear[ed] into the infinite text" only to be successfully "'born again' more purely as text" (Heywood 81). Language is what has "killed" her and language is what actually keeps making her "be born" and live again, in the stories and rumors about her that will continue to be told (including the present one), all pointing, in one way or another, to the resilience of the artistic powers of dreaming and imagination. Beyond

52 The fundamental linguistic (and thus also creative) operation at the base of the notion of "fame" emerges from its definition in the *OED*: "That which people say or tell; public report, common talk; a particular instance of this, a report, rumor."

the irony, her absurd death, together with the tale of her heroic suffering and alleged failure and defeat, may have the indirect effect of indefinitely renewing her (and her author's) reputation, of validating her final choices and affirmatively "ennobling her memory" (Frega 101).

In the last line of the novel, the narrator writes: "I morti non vedono i vivi." The courteous exchange between the caretaker and the ghost of the *massaia* regarding her forgotten "fazzolettino," suggests that a different kind of communication between "morti" and "vivi" is possible and that the former are still able to inspire, "be heard," have an influence and, especially, keep feeding the dreams and imagination of us all.

CHAPTER FOUR

How to Cope with the Bites of Modernity: Two Literary Recipes by Massimo Bontempelli and Luigi Pirandello

4.1 Learn to Play or Be Eaten: Massimo Bontempelli's *Gente nel tempo*

> Come si divertiranno gli uomini
> Quando avranno un coltello
> per tagliare l'Eternità
>
> — M. BONTEMPELLI, *Il Purosangue* 1919

> Vedi, tu sai tante cose, ma se non vuoi piú farmi paura, devi impararne una ancora: devi imparare a giocare.
>
> — M. BONTEMPELLI, *Nembo* 1935

Although a substantial amount of critical literature exists today on Massimo Bontempelli and his *oeuvre* as a modernist writer and intellectual, not many pages of criticism are dedicated to his 1937 "novecentista" novel *Gente nel tempo* (from now on, *Gente*).[1] In the most recent monographic studies on Bontempelli, their authors either did not feel the need – arguably because of their specific focus and goals – of including it in their analysis (i.e., Tempesti 1974; Saccone 1979; Urgnani 1991; Glielmo 1994) or, when

1 *Gente nel tempo* was written between 1935 and 1936 and it first appeared as a serial in the journal *Nuova Antologia*.

they did, they dismissed it quite rapidly, devoting minimal space to its discussion (see Airoldi Namer 1979; Cecchini 1986; Fontanella 1997).[2]

Granted, Bontempelli himself has notoriously been "un autore sostanzialmente 'rimosso' dalla critica letteraria italiana del secondo dopoguerra"[3] and, therefore, it should not be surprising that such a removal has affected also (and especially) one of his "più fortunati (!) romanzi,"[4] the one that, nonetheless, supposedly "segna il definitivo riconoscimento della maturità artistica dello scrittore."[5] As I am on the verge of taking a closer look at this text, one might at least speculate on some of the possible reasons for this sort of repression within a repression.

Perhaps, as Fontanella observes, it is due to the fact that *Gente* is a "romanzo un po' farraginoso, asfitticamente tirato per le lunghe e che risente di una scrittura a puntate"[6] or that "quanto è 'raccontato' in *Gente nel tempo* è banale [...] fatti mediocri [...] presi in prestito a volte da romanzi e da racconti già pubblicati dallo stesso Bontempelli."[7] Furthermore, perhaps this critical silence about the novel in the period after World War II may partially depend also on its somehow disturbing and unpopular subject matter. Writing a commentary on an apparently depressing narrative that includes a series of missed opportunities to create solid and rewarding personal relationships (betrayals, failed or impossible love stories); an episode of attempted rape; the death of an infant, the insanity of one of its main protagonists and, in short, the progressive, inevitable disintegration of an

2 A critical note by Marinella Mascia Galateria is contained in the 2001 edition of the novel (Milan: SE, 2001). Among the above mentioned scholars, Airoldi Namer and Cecchini are those who offer the most useful and insightful reflections for my purposes.

3 Roselena Glielmo, *La traversata dell'ironia: studi su Massimo Bontempelli* (Napoli: Guida, 1994), 7.

4 Elena Urgnani, *Sogni e visioni: Massimo Bontempelli fra surrealismo e futurismo* (Ravenna: Longo 1991), 14.

5 Carlo Cecchini, *Avanguardia mito e ideologia: Massimo Bontempelli tra futurismo e fascismo* (Roma: Il Ventaglio, 1986), 152–3.

6 Luigi Fontanella, *Storia di Bontempelli: tra i sofismi della ragione e le irruzioni dell'immaginazione* (Ravenna: Longo, 1997), 90.

7 Fulvia Airoldi Namer, *Massimo Bontempelli* (Milano: Mursia, 1979), 121–2.

apparently normal bourgeois family until its extinction over time,[8] must not have appeared particularly inviting, edifying, or fashionable to a number of Italian literary critics (regardless of their political orientation) during the booming, optimistic, and increasingly consumerist post-war decades. In other words, one may reasonably expect that *Gente* did not attract many critical readers in years when the attempt to modernize the nation, reconstruct (rather than de-construct) the idea of the family, and reassure the Italian "gente" about their rosy future and prospects as active (re)producers and consumers were major socio-political and cultural concerns.

Following a similar logic, one wonders also how fascism or its intellectual elites and ideological apparatus, which notoriously used the family as a "strumento manipolatore" in order to achieve a "fascistizzazione totale della vita italiana" (Cecchini 112) and supported fascism's propaganda and myths of rebirth and immortality with the notes of anthems such as "Giovinezza,"[9] must have reacted to the appearance of Bontempelli's novel. And what kind of response could there have been, other than a relative silence, to a text that allusively opens with the death of a dictatorial "Mussolinesque" "Gran Vecchia," suggestively characterized by a "mento potente" and a big, highlighted "testa"[10] and proceeds by displaying the regular disappearance of all her descendants, dismantling in the process any apparent hope of renewal or possibility of other than an ephemeral and painful permanence in the world?

Elena Urgnani observes that "Le opere di Bontempelli evit[a]no il più possible di occuparsi di politica." (8). Her statement undoubtedly

8 It is probably a sign of Bontempelli's ironic attitude that the family's name is "Medici" and that all its members are victims of the same, incurable time-induced "sickness."

9 "Giovinezza" (words by Salvator Gotta (1924) and music by Giuseppe Blanc (1909)) notoriously extols the eternal youth of the fascist party and its brilliant future. Here are its initial lyrics with their emphasis on ideas of immortality and rejuvenation: "Salve o popolo d'eroi/Salve o patria immortale/Son rinati i figli tuoi/Con la fe' nell'ideale/Il valor dei tuoi guerrieri,/La virtù dei pionieri/La vision dell'Alighieri/Oggi brilla in tutti i cuor."

10 I am obviously suggesting that the "Gran Vecchia" resembles the "Duce" both in her aggressive and commanding manners and her physical appearance.

simplifies the task of those, among Bontempelli's scholars, who have to deal with the well-known inconsistencies and ambiguities of his political position from the mid twenties to the late thirties (and beyond). However, if one recalls that the fracture between Bontempelli and the fascist regime began to widen in the years after 1934, it is tempting to suggest that *Gente* could (also) contain a veiled criticism of the "tempo" of the regime itself (i.e., of the contingent "storia in atto" in the period),[11] of fascism's absurd myth of timeless, youthful self-perpetuation, and of its eventually deleterious and stifling effects on whoever belongs to its larger, social "family"; a family which, in its extended meaning, cannot exclude the figure of the contemporary artist and intellectual.

On the one hand, fascism's self-serving myth-making tendency, its eagerness to escape from historical contingency and its problems, could have beem appealing to and instinctively shared by Bontempelli. His own rejection of historicism, his proposal to "ricostruire il tempo e lo spazio" and to create "modern myths" and, finally, his "magical" "riduzione mitologica della storia"[12] are all gestures that, in a general sense, are elements of a fascist vision of the world. On the other, however, history's inescapable resurfacing in the reality of the regime itself (i.e., fascism's' "progressive" belief in improving its own "apparati di stato," Urgnani [9]), not to mention fascism's narrow, propagandistic understanding of art and the artist were always problematic issues for our writer and, as a consequence, among the sources of his intellectual unease toward and criticism of the regime.[13]

11 See (Airoldi Namer 1979, 14/25).

12 Antonio Saccone, *Massimo Bontempelli: il mito del '900* (Napoli: Liguori, 1979), 99.

13 Several of Bontempelli's essays exemplify this divided position. So, for example, while in his essay "Personali ma non troppo" (1929) he brings together mythology and (fascist) politics in the same paragraph: "La poesia coincide all'origine con l'invenzione del mito e della favola [...] La mia antica adesione al fascismo si deve soprattutto al fatto che lo considerai come un franco primitivismo politico, cancellate allegramente e d'un sol tratto tutte le esperienze delle consumate politiche che lo avevano preceduto" in *L'avventura novecentista* (Firenze: Vallecchi, 1974), 187–8, in "Protezione" (1936), writing about the relationship between art and politics, he complains that "da alcuni anni ci si affanna a volere che l'arte assuma l'ufficio di propaganda politica dichiarata" (Bontempelli 1974, 117). A couple of years later, in

This emphasis on the presence of a politically charged message in *Gente* is not because I wish to go over Bontempelli's controversial relationship with fascism once again. Rather, it is because its individuation confirms, from yet another perspective, the writer's fundamental distrust and rejection of history ("nella storia [...] c'e' soltanto la morte," Cecchini [149]). Most importantly, it reintroduces a motif – the one of the alienated subject/artist who, oppressed by the ruling political power and the "historical," capitalist, industrial cycles of production and standardization, feels in constant danger of being "devoured" by the era in which he/she operates – that not only summarizes a familiar way of describing the experience of modernity, but was a crucial concern of Bontempelli from the beginning of his writing career.[14]

In this sense, *Gente*, as its title suggests, represents an example of Bontempelli's thematic obsession with time and history.[15] The novel can thus be interpreted as a reflection on an essentially Nietzschian "sickness" of history and on attempting to escape from it, as Bontempelli wrote,

"Le rane chiedono tanti re" (1938), he argues against the need for an artist to have "direttive e [...] disposizioni per sapere quel che è morale" and concludes by claiming his creative freedom: "preferirei mi si togliesse addirittura di mano la penna piuttosto che sentirmi uno dietro le spalle intanto che scrivo" (Bontempelli 1974, 216).

14 *La vita operosa* (1920), for example, "sottolinea soltanto il fatto che nella società industriale non c'è posto per lo scrittore, per il letterato tradizionale, che il suo lavoro non è produttivo, non è merce ..." (Cecchini 42). This idea is reiterated in his 1935 essay "Tipi che scompaiono:" "Molto indebolita dal 1870 alla fine del secolo scorso, fatta rarissima nei primi quindici anni di questo, al ritorno dalla guerra abbiamo trovata quasi del tutto scomparsa una categoria, una figura, che aveva vissuto per secoli una vita ininterrotta, e in certi periodi abbastanza fervida. È la figura del 'poeta.'" (Bontempelli 1974, 109).

15 Interestingly, more or less during the same period, Giuseppe Ungaretti too in *Sentimento del tempo* ("his fall out of Eden into time and history"), shares a similar obsession. See Margaret Brose, "Giuseppe Ungaretti's *Sentimento del Tempo*: Baroque Rome and the Experience of Time," *Pacific Coast Philology*, 21, 1–2 (November 1986), 65–72. See for example his "La fine di Crono" (1925): L'ora impaurita/In grembo al firmamento/Erra strana./Una fuligine/Lilla corona i monti,/Fu l'ultimo grido a smarrirsi./Penelopi innumeri, astri/Vi riabbraccia il Signore!/(Ah, cecità!/Frana delle notti)/E riporge l'Olimpo,/Fiore eterno di sonno.

"con tutti i mezzi, dai più brutali ai più morbidi" (Bontempelli 1974, 27). Accordingly, as *Gente* displays the nefarious effects (arguably, both on the doomed Medici family and on Italian artistic culture) that are likely to result if the "secolo ventesimo" fails to accomplish that "compito urgente e preciso […] la ricostruzione del tempo e dello spazio" (Bontempelli 1974, 9), it also epitomizes the poetics of *Novecentismo*, in which the principle of "magical realism" and the correlated rediscovery of the salvific, inspirational power of myths constituted the movement's core elements.[16]

Now, while *Gente* is remembered as a manifestation of such a poetics and of the mythical quality of Bontempelli's fiction, the myth that lies behind the narrative and inspired the novel has barely been mentioned in a critical context and, subsequently, the myth's potential hermeneutical implications have never been analyzed in depth. The main objective of this section is to fill this apparent void especially in the light of Bontempelli's overall aesthetic project and his ideas on artistic creativity *vis à vis* history and time. Ultimately, I would like to suggest that the novel, while possibly pointing to the "relentless devouring of human time" (Michaud 42) operated by the monsters of modern society (among which industrialism and capitalism play a decisive role), may hint at a way through which such "human time" may not only be "magically reconstructed" but also salvaged and re-experienced.

As far as I have been able to establish, Fulvia Airoldi Namer is the only scholar who has not found it irrelevant or too obvious to note that the "Gran Vecchia" is the "incarnazione femminile e materna di Crono che divora i suoi figli" (Airoldi Namer 1979, 122) and who has pointed out that the classical, Greek myth of Kronos – the Titan god who swallowed each of his children as soon as they were born to preempt the prophecy that one of his own sons would overcome him – lies at the imaginative core of the novel. The tragic story of the Medici family and, above all, of

16 "Il realismo magico, alla ricerca della dimensione mitica assoluta nel mondo contemporaneo […] esprime essenzialmente il bisogno di sfuggire alla presa della storia e quindi a ogni verosimiglianza." Qtd. in Piero Taravacci. "Il realismo magico di Bontempelli," *Trimestre* 13 (1980), 217–37.

the two sisters Nora and Dirce, appears to be, at its most basic level, a fictional representation of the tyranny of time, of the "rule of Kronos" and its all-devouring features.

However, the classical myth includes a *pars destruens* (i.e., the swallowing of the children), and what may perhaps be interpreted as a *pars costruens* since, in the end Zeus, as the only saved offspring, forced Kronos to disgorge his swallowed children (Zeus' siblings) who, once recreated, proceeded together to defeat the Titans, including Kronos, in a ten-year war and relegated them to the pit of Tartaros.

The kind of time Kronos symbolizes, defined by its rigid, quantitative, fixed and measurable qualities (the five-year intervals), not only brings it closer to the category of space (the gloomy "villa detta la Coronata" lurks always in the background) but emblematizes the all-consuming idols of modernity. It also evokes by contrast and default Kairos, the other, alternative and yet complementary word the Greeks used to define time and to which, in turn, Bontempelli alludes to whenever he suggests that it is only through a *comprehensive* recuperation of the mythical dimension of temporality that we may have a "rapporto organico e non alienante [...] col proprio tempo" (Cecchini 51). While Kronos refers to chronological or sequential time, Kairos represents its qualitative nature; it is the "fullness of time; the propitious moment for the performance of an action or the coming into being of a new state" or, additionally, "a time in between, a moment of undetermined period in which something special may happen" (*OED*). The latter is, apparently, the kind of time that Mircea Eliade had in mind when he conceived his original opposition: on the one hand a "profane" time and, on the other, a (kairotic) "creative," "regenerative," "sacred," "mythical" one, punctuated by rituals and, ultimately, a way to access eternity.[17]

If it is true, as I shall try to demonstrate in the following pages, that the novel displays both the explicit, dyspeptic scenario incarnated by "profane"

17 Mircea Eliade, *Patterns in Comparative Religion*, trans. by R. Sheed (Lincoln and London: University of Nebraska Press, 1996). On Eliade's notion of "archetipal repetition" as conducive to an a-historical "sacred time" see Cecchini (26–8).

Kronos and the implicit allusion to a "sacred" and regenerative time represented by Kairos, that display could be a sign that *Gente*, rather than being merely characterized by its updated, modern, nihilistic dramatization of Kronos' rigid destructiveness (and, by extension, of the mere negativity of history), may also point to an eventual "re-constitutive" phase and redemptive possibility. In this scenario, Kronos as emblem of both capital's greed and time as a divisible, measurable entity would ultimately be destroyed (or "tagliato") and substituted by (or "reconstructed as") Kairos: time as "opportune moment," in which change, "victory" and, thus, also artistic creativity become feasible events.

The "ricostruzione" and, implicitly, the escape from a paralyzing and stifling history may thus begin from a radical transformation of how time itself is perceived and experienced. I would like to suggest that among Bontempelli's concerns in *Gente*, the depiction of the negative effects of Kronos is joined by the evocation and affirmation of the alternative, redemptive, and creative possibility associated with Kairos, or with a form of "human time" which, in a utopian perspective, does not run the risk of being devoured by the monsters of modernity.

While in *Gente* Bontempelli is "having fun" (or "plays") (à la Palazzeschi) at literally and literarily "tagliare" Kronos with the "coltello" of his imagination, he also may be laying in the same novel the foundation for that "spiritual geometry" which he deems necessary for the "recovery of the individual" (Bontempelli 1974, 9–10). And he seems to be doing so by repeatedly alluding to an imaginative "kairotic" experience where man may re-live a different sort of time (and, as a consequence, be "recovered") and artistic creativity may expect to reacquire a pre-eminent, active role. I would like to argue that "play" in all its ramifications, constitutes this sort of experience.

As early as the opening pages of the novel, a few hints suggest the possibility that such a "recovery of the individual" already (although paradoxically) begins with the disappearance of the Gran Vecchia. For example, while the family is reunited around her deathbed on the evening before her actual death, her voice is said to have a salvific effect: "La Gran Vecchia rise forte, a quel ridere nacque nell'animo di tutti un gran pànico, forse si buttavano dalla finestra ma lei ancora una volta li salvò riprendendo di colpo a parlare." (602).[18]

Once the Gran Vecchia is expeditiously sent to Tartaros after having pronounced her prophecy "nessuno di voi morirà vecchio" (602),[19] the potential of experiencing a different sort of time (Kairos) may open up for both the "gente" of Colonna and – more problematically – her descendants. In its first and most immediate manifestation, this experience coincides with the "mythical," ritual time of the "festa." To put it with Cecchini's Eliade-inspired words: "La morte della Gran Vecchia finisce per diventare [...] un vero e proprio simbolo di salvazione [...] Nel binomio morte-festa risiede insomma l'annullamento della creazione, ma, contemporaneamente, la continua possibilità di rigenerarla" (Cecchini 157–8). From this perspective, the activities that define such a feast are particularly revealing:

> I ragazzini in piazza giocavano ai funerali; la fanciulla scelta a figurare la Gran Vecchia era felice d'imbiancarsi la faccia col gesso e mettersi un cuffione di carta, poi sdraiata a occhi chiusi sopra un vecchio battente d'armadio farsi portare a spalle dai più forti, e gli altri dietro cantando con voci nasali. Quando fu la sera molti invece di rincasare desinarono all'osteria [...] in molte case si rimase a ballare nell'atrio al suono d'un organetto (610–11).

The playful game performed by the children of Colonna, a mock-theatrical spectacle representing the funeral of the Gran Vecchia, and the locals' communal dinner outdoor followed by dances and music at home are the kind of primal, archetypical events whose timely appearance sanctions the liberation from the Gran Vecchia as emblem of the "rule of Kronos" (and, thus, from the mutable passing of time),[20] and the simultaneous longing

18 Similarly, the death of Nora's baby, Fausto, will later save the two sisters: "Nora, che aspettando Fausto diceva a Dirce 'ecco lui ci salva nascendo,' non aveva ancora pensato che invece lui le aveva salvate morendo." (736).

19 Cecchini already noted that the curse of the Gran Vecchia "al di là di tutte le possibili rotture, [è] un rassicurante esorcismo contro la storia, contro il tempo che nell'arco di venticinque anni rimarrà [...] uguale a se stesso" (Cecchini 157).

20 Not surprisingly, after her death, the narrator observes: "Non c'era piú ragione che il tempo passasse, che quella notte idiota arrivasse a una fine." (606).

for a fuller, eternal, "kairotic" "perduto paradiso delle forme fisse" (Airoldi Namer 1979, 13).

A few pages later, as the narration begins to focus on Silvano, the next in line to die as the son of the Gran Vecchia, and on his daughters, Dirce and Nora, we encounter another sign that the matriarch's death is associated with a regenerative pattern:

> – Babbo, babbo – e una da un lato l'altra dall'altro balzarono sulle sue ginocchia [...] – Sai? – disse Dirce – siamo arrivate giú alla fontanella del capelvenere che s'era seccata, invece ora ricomincia a dar acqua.
>
> – Andiamoci – gridò Nora con entusiasmo – noi lí facciamo sempre il gioco del precipizio (616).

No special exegetical skills are needed to note that water, a primary, life-bearing element and a recurrent symbol of the eternal in Bontempelli's work,[21] begins to gush out of the dry fountain in synchrony with the death of the Gran Vecchia. More interesting to note, however, especially if we recall Johan Huizinga's remarks on the peculiarities of play and its contiguity with archetypes,[22] is that a game, in this case an eminently "superfluous" and "disinterested" activity invented by the two girls, is linked once again to this site and moment of renewal. Like all play, the one that involves the girls is one that is "in-ordinary" and, most appropriately for this context, that "stands outside the immediate satisfaction of wants and appetites [...] *it interrupts the appetitive process*" (italics mine).[23] As such, it can be said to interrupt (even though temporarily) the devouring pattern of Kronos and to belong to the special, creative realm of Kairos.[24] Play, in all its variations

21 See, among others, his short story "L'acqua."

22 "The great archetypal activities of human society are all permeated with play from the start. Take language [...] Or take myth [...] Or, finally, let us take ritual" (Huizinga 4–5).

23 Johan Huizinga, *Homo Ludens. A Study of the Play-Element in Culture* (Boston: The Beacon Press 1950), 9.

24 Georges Bataille's observations in *The Accursed Share* seem particularly pertinent in these circumstances. For him play can break man's slavery to "production, accumulation and utility." It "can disrupt the operations of the limited economy by consuming,

including, by extension, artistic creativity may thus constitute both the main instrument with which to free oneself from the destructive voracity of historical time, and the ultimate object of the writer's deepest longing.[25]

When Silvano rejects his daughters' invitation to join them in their game ("– Allora il primo che arriva – spiegò Dirce – ha diritto di buttare un po' d'acqua addosso agli altri [...] – No, care. Andate voi. Io non posso" [616]), he "misses his chance;"[26] he fails to seize an "opportune moment" that would momentarily "suspend time." His refusal to play in conjunction with his missed contact with water symbolically marks his distance from Kairos and anticipates his imminent falling back into the "precipice" or, else, into the jaws of Kronos.

Granted, Silvano is said to play too: "Tu – disse Nora – sai giocare solamente coi libri." (617). However, Bontempelli seems to imply, when books are intended merely as objects to stack and catalogue, they cannot provide any real fun. Silvano's game, to put it with Huizinga, appears to lack "the formal and essential characteristics of play" (Huizinga 18) and, therefore, it is never able to "transport [him] to another world," or to detach him from his "ordinary life" and, as a consequence, from the "sickness" of

---

by squandering resources, energy, and time. Play can disrupt the ethos of utility and the enslavement to useful ends." Qtd. in Dorothy Holland, "Bodies at Play: A General Economy of Performance" in Shannon Winnubst, ed., *Reading Bataille Now* (Bloomington and Indianapolis: Indiana University Press, 2007), 197–219, (207). So, it can evidently also break Kronos' rule.

25 In his "Discorso ai littoriali sportivi" Bontempelli himself establishes a connection between "sport" and "arte": "Se mi permetteste di fare un momento della astrazione di colore filosofico, potrei dire che lo sport è *azione disinteressata*. Per questo rispetto esso verrebbe a trovarsi in un allineamento delle operazioni umane, accanto all'arte, che è *comunicazione disinteressata*" (Bontempelli 1974, 99). Additionally, as Eugene Fink observed: "In the autonomy of play action there appears a possibility of human timelessness in time. Time is then experienced, not as a precipitate rush of successive moments, bur rather as the one full moment that is, so to speak, a glimpse of eternity." Eugene Fink, "The Oasis of Happiness: Towards an Ontology of Play," *Yale French Studies* 41, (1968), 19–30, (21).

26 The motif of missing one's chance and opportunity or failing to seize the right time affects also Maurizio who thinks: "Avrei potuto proporle di sposarla [Vittoria], non sono stato pronto; c'è tempo" (647).

temporality. Although Silvano is passionate, his involvement with books in general, and the "preziosi fascicoli" received as a gift from Abate Clementi in particular, is that of a solitary collector. It is the pastime of a quintessentially hungry, bourgeois, capitalist bibliophile who is merely interested in cataloguing and accumulating (or "gorging on") them as objects for their rarity, external appearance and value ("era opportuno farli rilegare" [623]) with the precision of an accountant: "Non leggeva penetrando il senso, leggeva da bibliofilo; tutt'al piú seguiva il correre d'un periodo [...] Sfogliò il volume: non c'è dubbio, mancano le pagine da 561 a 624, sessantaquattro pagine, quattro sedicesimi" [638–9]).

The "greedy" "book-game" Silvano practices is not only far from real play in the sense individuated by Huizinga, that is, a social "activity connected with no material interest, and no profit can be gained by it" (Huizinga 13), but also has the effect of assimilating the character of Silvano (and, together with him, of the other avid collector, Abate Clementi, "l'*alter ego* razionalizzante del narratore" [Airoldi Namer 1979, 124]) to the figure of a dull, pedantic and obsessed individual, one, in short – Bontempelli seems to suggest – who, like many other contemporary representatives of Italian letters and cultural life, is not able to "[penetrare] il senso" and thus is in dire need of being "recovered."

Silvano's search for the missing pages (notably, his last activity before his death), acquires an additional layer of meaning in this light. Perhaps it is because he is "non-playfully" obsessed by the desire of finding and laboriously reordering this exact amount of pages – which, in their chronological sequence, seem a physical manifestation of the quantitative and measurable qualities of time – that, in turn, Kronos, disguised as a ghastly Gran Vecchia (here a classic manifestation of a Lacanian "devouring mother"), catches up with him and swallows him down: "Allora dal fondo dell'ombra, dal fondo del silenzio, il volto della Gran Vecchia lampeggiò, e Silvano non provò piú niente" (643).

It is, however, the Medici's home, Villa Coronata which symbolizes best the overlapping of the old, quantitative spatial and temporal dimensions. This is probably why real play cannot initially occur within its walls. If Nora and Dirce are able to play in the garden of the villa in a dynamic and inventive way ("[il] gioco [...] era inseguirsi con la legge che l'inseguitore non deve passare davanti a certi alberi, né l'inseguito a certi altri"), when they move indoors the girls "diventavano inerti; l'austerità dell'interno, ch'era rimasta immutata, non favoriva lo spirito del gioco" (621).[27]

Only when Maurizio (a close friend of the family) brings inside a "mazzo di carte (in quella casa non se n'erano mai viste) e insegnò loro alcuni giochi facili" (621), can the two sisters feel alive, passionate, and enthusiastic ("Per Dirce e Nora fu un nuovo mondo rivelato, come quando nell'infanzia avevano scoperto i tre regni della natura" [621]). The mysterious symbols and pictures on the cards stimulate the young girls' interest and creativity. Dirce is eager to be taught more adult card-games, a desire revealing her equally adult and bourgeois, business-like skills of "memoria rapida e molta astuzia." Nora has more fun playing with her newly stimulated imagination than with the games she recently learned: "fantasticava a lungo la notte al buio prima di addormentarsi, mescolando tutti quei personaggi in avventure turbinose" (621–2). This latter kind of card-playing is an activity coterminous with the one of a story-teller, a figure who is able to (be transported) and transport her/his listener to other spatial-temporal dimensions or to forget momentarily the rigid tyranny of a quantitative temporality (Kronos) and privilege instead a qualitative one (Kairos). This sort of play confirms once more its eminently aesthetic character, its coincidence with the realm of the imagination and its possession of a sort of prophylactic value. It constitutes one of the barriers against the insatiable plague represented by Kronos. The ideal world full of "avventure turbinose" created by Nora and, at another level, the parallel, magic world of the novel written by Bontempelli thus form a similar kind of "heterotopic" space momentarily suspended out of history and, simultaneously, the necessary game one needs to play in order to begin any attempt at recovering the individual in the real world.

27 Only when the two sisters move to Milan, in their first, small and yet fashionable apartment, are they able to play indoors: "questo fu per loro un divertimento, pensavano d'essere diventate bambole, giocavano di stanza in stanza come dieci anni prima d'albero in albero alla Coronata" (668). Focusing on their own space and on their own game momentarily relegates time to the background.

The prophylactic (and therapeutic) potential that play, including here the darker, more shadowy sides of human enjoyment, may have as a defense against Kronos' voracity and the usual, "purposive structure of our life" (Fink 22) emerges again when Dirce refuses to join Maurizio and her sister on their trip to purchase an automobile:

> – No – rispose – ho promesso, dopo la messa, di passare questa domenica in casa. [...]
> – Promesso, a chi?
> – Dirce rispose con compunzione:
> – Un fioretto.
> Dovettero spiegare a Maurizio che le monache insegnano a fare i fioretti, atti di rinunzia spontanea a qualche piccolo piacere per acquistarsi grazia. Dirce l'aveva presa molto sul serio: una sera si inibiva di mangiare le frutta, un giorno s'imponeva di restare seduta sul suo banco tutto il tempo della ricreazione, e simili.

The "fioretto," Dirce's little game of renunciation and sacrifice is a sort of protective strategy that, by interrupting her own appetites, aims to make her less appealing and "digestible" for Kronos. It aims to interrupt his destructive appetite and, at the same time, to provide her acquisition of "grazia" here in the sense of literally "being spared." As she will later cruelly remind her sister Nora, her decision to remain home to fast has possibly saved her from experiencing Petronio's attempted rape: "... se ieri mattina stavi a casa come me, per fioretto, non ti trovavi in quel bel pasticcio" (652).[28]

It is worth noting that references to both sacrifice and alimentation reappear in close relation to Nora's brief re-living, together with her new, older and wealthier neighbor in Colonna, Lando Della Volpe, of her terrible experience with Petronio. When we first encounter Lando, he is reflecting on the possibility, mentioned by his father, of a marriage of convenience with Dirce, which would help increase the social and political capital of his own family: "'non c'è nessun bisogno di sacrificarsi, per farsi padrone del paese [...] Certo, sacrificarmi. Sarei ridicolo. Non mi sono mai piaciute

28 Tangentially, the fact that Dirce's "strano [...] contegno" appears in relation to the imminent trip to buy an automobile, one of the most powerful symbols of industrialization, reinforces the parallelism between the allegorical, devouring image of Kronos and any other equally dangerous manifestations of the modern times.

le ragazzine. Bel gusto.' […] uscí di camera canticchiando 'bel gusto' su toni varii" (655).

The narrator's emphasis on the notion of "gusto" and the subsequent implication that young girls are something one may metaphorically taste and eat seem to make of Lando's character not only a greedy landowner and calculating politician but also another personification of Kronos' devouring pattern.

Lando's refusal to sacrifice (significantly, in order to avoid being perceived as "ridicolo" or "made game of") corresponds to an intensification (rather than an interruption, in opposition to Dirce and her fasting) of his appetites. In this light, it is not surprising that the paragraphs that immediately follow his introduction in the narration deal with food and eating, as they describe the overabundant breakfast he is offered by a local farmer:

> La seconda di queste giornate fu quasi tutta presa da una colazione che gli offersero in un podere della pianura: colazione di ciclopi, che cominciò alle undici e finí alle cinque. […] Lando incontrò la brigata degli adolescenti al campo del tennis. Lo burlarono per il suo volto rosso e gli occhi lucidi, lui li fece ridere descrivendo il formidabile pasto (655).

The mythological reference to the Cyclops,[29] the precise, chronological indication of the duration of the "formidable meal" and, finally, Lando's own, altered physical appearance marked by the signs of overeating are further indications of his closeness to Kronos and his predatory, insatiable behavior. It seems also relevant that Lando's first encounter with Nora happens while she is playing a tennis game – while Nora is in a sort of "safe-zone", arguably protected from Lando's/Kronos' appetites (real and symbolic) by the kind of playful activity she is currently engaging in:

> [Lando] Guardò ancora Nora, che aveva ripreso la partita […] Si mise a passeggiare lungo il campo. Una palla gli arrivò tra i piedi. Si chinò a raccoglierla, intanto Nora arrivava in corsa: "Bravo, che è riuscito a chinarsi." Prese dalle mani di lui la palla e di nuovo corse via. […] Nora giocava con impeto, si lanciava, aveva veemenze che esaltavano tutte le sue forme … (656).

29 In Hesiod's *Theogony* the Cyclops (Brontes, Steropes, and Arges) were the sons of Uranus (Sky) and Gaia (Earth). And so, by the way, was Cronus/Kronos.

As soon as the game is over, however, Lando makes his move. He approaches Nora, puts his arm under hers, and then – while she naively and playfully makes fun of him, alluding again to his age and then to his eating habits ("Si appoggi, poverino – disse Nora – ma vede che vergogna darsi ai piaceri della gola?" [656]) – he pulls her towards him and kisses her on the mouth. Nora herself, evidently, represents nothing other than one such tasty "piacer[e]" for him. She is the privileged object of his desire/hunger, and when the game is over it is only her firm reaction that saves her from being metaphorically eaten. This incident is quickly dismissed by Lando, "tornato in sé," as an innocent "scherzo." But from Nora's perspective, what has happened is as far as it can possibly be from any joke, trick, or amusing game, since it reminds her of Petronio's bestial aggression. As the language confirms, she now perceives Lando's actions precisely in terms of an oral aggression and an attempt on his part to feed on her: "in quell'attimo lei d'improvviso s'era veduti sulla faccia i denti bianchi di Petronio, s'era sentito addosso il corpo di Petronio" (657).

This combination of the alimentary, the sexual, and the ludic spheres recurs elsewhere in the novel in a similarly subtle and articulated fashion. The seduction game between Nora and Dario, for example, develops against the background of the dinner-at-home they missed because of their love-induced tardiness and of a game of poker between Dirce, Giuliano and two other guests.

In this episode, the eminently kairotic dimension in which Nora and Dario move (they happily wander around the town; they are indifferent to time constraints and, in short, they are forgetful "che intorno c'è il mondo" [682]) is juxtaposed with the rigid, aggressive, and ravenous behavior of Dirce who, paradoxically, ends up as a grotesque personification of Kronos' more prominent characteristics (i.e., its quantitative, chronological concern and its aggressive voracity):

> Erano le nove, un'ora di ritardo, e Dirce è terribilmente precisa in queste cose. [...] Dirce, Giuliano e Narcisa avevano aspettato mezz'ora prima di mettersi a tavola. Ora han finito [...] Alla tavola tre posti si vedevano sfatti e ignominosi come dappertutto ove l'uomo ha sostato a nutrirsi. Il posto di Nora era immacolato (679–80).

First, Dirce's indirect association with the nutritional behavior of a man momentarily marginalizes her female gender and further supports her assimilation by Kronos' canonical masculine features.

Additionally, although Nora tells Zelinda (the maid) that she and Dario are hungry ("Noi abbiamo fame; Zelinda, porta qua tutti gli avanzi" [680]) and eventually will nibble something in the kitchen, the couple does not participate in the dinner. The dinner, therefore, seems just a pretext to elicit the narrator's depiction of the negative effects of the voracity of Dirce, Giuliano, and Narcisa. These characters, in fact, are presented here according to their relation with a Kronos-like time and its devastating effects on the surrounding space.[30] Nora's "immaculate" spot, therefore, not only alludes to her empty stomach, lack of food intake and, arguably, to her alternative, less structured alimentary habit ("Mangiamo qui, mangiare in cucina è una delle cose piú divertenti e non me la lasciano fare mai" [680]), but also to the gleeful, special (kairotic) period that she is experiencing thanks to her infatuation with Dario.

Along these lines, Nora's refusal to join the poker game with Dirce ("Noi – dichiarò Nora – non stiamo né alla seduta né al poker" [680]) emphasizes that the latter, as a "giocatrice attenta e avida" (681), represents a Kronos-like character and, therefore, an imminent danger for Nora and her finances, both at risk of being easily "swallowed up." It also indicates that poker's qualities, as a game of chance, are quite different from those of play as I have tried to delineate them. If play as represented so far by Bontempelli may be defined as *paidia*, it belongs to the kairotic dimension, performing a prophylactic role against Kronos' all-consuming hunger and offeiring the possibility of experiencing "human timelessness in time"[31] (Fink 21), then poker (as a game belonging to the category of *alea*) has

30 One could perhaps imagine that Bontempelli's "ricostruzione" aims precisely at transforming an "unmade and disgraceful" space into an "immaculate" one.

31 Appropriately, when in Sirmione, one of the stops in Nora's and Dario's love-journey throughout Italy, the narrator observes that: "Scendendo all'albergo, pareva loro di entrare in una immutabile eternità" (684).

to do with profit and material interest and, as a consequence, it increases (rather than interrupts) any appetitive (or consuming) process.[32]

When taken together as gestures of abstinence, Nora's own alimentary and gaming decisions indicate that her joyful (*paidetic*) attitude acquired from her flirting with Dario may actually play a role in preventing her from becoming some kind of food herself. By skipping the dinner and avoiding being reduced to a passive commodity for someone else's gain in the poker game ("Dirce [...] riusciva a non perdere quasi mai e quando perdeva non pagava" [681]), she becomes temporarily "inedible," escaping the danger of being consumed. At the same time, Dirce/Kronos, once deprived of the possibility of financially consuming her sister, can only manifest her aggressiveness (and jealousy of Nora's relationship) orally, with a verbal attack: "– Nora – strillò – vergogna vergogna [...] si voltò [...] come un serpe" (682).[33]

Dirce's active and profit-oriented involvement with card-games ironically (and conversely) returns to haunt her when, back in Colonna to talk with Abate Clementi about the five-year-curse, she happens to enter the local "locanda del *Gallo*." Here she is aghast at realizing that she and Nora are themselves the unaware, passive protagonists of another, very particular "gioco di carte," a betting game organized by the locals who speculate and gamble on the time of their death, as if it were a horse-race:

> Era una lavagna appesa al muro: il piano nero era diviso verticalmente con una grossa riga bianca in due campi. In alto d'uno dei due campi era scritta [...] una gran D maiuscola; in alto all'altro una N. Sia sotto la D, sia sotto la N, erano numeri mezzo cancellati. [...] Entrò l'oste. [...] – Che cosa vuol dire quello – gli domandò Dirce indicando la lavagna. [...] – Conti, vecchi conti di gioco, qui fa più caldo e la sera

32 I am again borrowing the terms *paidia* and *alea* from Callois' classification of games. If the former is characterized by "free improvisation, and carefree gaiety," then in the latter "the player is entirely passive [...] He risks his stake [...] In one instant, winnings may be wiped out." Roger Callois, *Man, Plays and Games*, translated by Meyer Barash (New York: The Free Press, 19), 13–17. Among Callois' other categories are *agôn* (competition), *mimicry* (simulation) and *ilinx* (vertigo).

33 The alliterative emphasis on the "growling" consonant "r" in these lines seems to further emphasize my point about Dirce/Kronos bestial voracity.

> vengono qui a giocare, a carte, giochi del paese s'intende, non li so bene, io non so fare nessun gioco, sto a guardare; e loro, ecco segnano i numeri dei punti là, col gesso: tutti giochi di carte ... (719).

That the place where this game of chance takes place is a *locanda*, a site where people eat and drink ("v'erano cinque o sei tavolini apparecchiati, ognuno con due o tre commensali, e lungo la parete una tavola lunga con tanti altri seduti intorno a mangiare e bere e parlar forte" [718]), once again contaminates the ludic with the alimentary motif. While emphasizing that *alea* and appetite are interconnected, linked as they are by a rhetoric of material gain and consumption, and that play as *paidia* contributes instead to the interruption of such pattern of voracity, the episode in the *locanda* reminds us that Dirce's and Nora's most pressing concern – a crucial part of their desperate attempts to postpone their final departure – may well be to avoid becoming food themselves. To put it differently, they wish to avoid becoming consumable commodities, merely useful for someone else's gain.

The novel reaches its poignant conclusion precisely with a final reference to issues of commodification, consumption, and play as "mimicry." After Fausto's death and Nora's subsequent suicidal sacrifice for the sake of her sister, the final page of the novel describes an enigmatic scene in which Dirce, the sole survivor of the Medici family, stars as the progressively crazier protagonist. Rather than answering the impossible question of Abate Clementi ("Cinque anni. Che cosa ne farai, Dirce?" [750]):

> Dirce s'alzò in piedi, spiccò un salto, e si mise a ridere [...] Dirce a salti quasi ballando, e sempre ridendo forte, traversò la piazza, entrò in una strada ov'era la chiesa, sedé sui gradini della chiesa, si strappò il fazzoletto dalla testa ch'era diventata tutta bianca come quella dell'abate Clementi. Aguzzò lo sguardo e vedendo gente spuntare e avviarsi alla chiesa smise di colpo di ridere, con un lampo malizioso negli occhi chinò il capo, si strinse nelle spalle, tese una mano e cominciò a chiedere l'elemosina (750).

Dirce's final begging act has been generally interpreted either as a desperate, symbolic plea for more years to live or, alternatively, as her "abbandono totale al divenire" (Airoldi Namer 1979, 124). However, her jumping, immoderate laughter, and dancing, culminating in the even more theatrical gesture of pulling her headscarf away, reveal a different, transformed self and allow me to look at her as a kind of mad clown, a twisted manifestation of a *saltimbanco*. Her gestures resemble those of someone performing an act, who is literally (perhaps even consciously) playing the part of the beggar as she sees people approaching the church, changing her appearance and physical stance for the occasion (and in the category of "mimicry" Callois significantly lists "Masks" and "Disguises").

What her "malicious glance" initiates, then, is much more than a simple request for spare change or, figuratively, for more time. Whether inspired by reason or folly, her histrionic "begging game" is of a very special sort. What it accomplishes, in fact, is that by removing her from the productive cycle and economic dynamics of a bourgeois, capitalist society and placing her, instead, at its antipodes as a beggar, her "begging game" modifies her status of "consumable commodity" or "food for Kronos." As long as she begs (or plays her game), she can avoid being devoured by Kronos and, thus being consumed by the other "monsters of modernity" associated with him. From a symbolic perspective, moreover, it seems evident that her decision to renounce material possessions and to rely only on others' voluntary gifts, is not that far from the "starving", ascetic behavior she manifested previously with her "fioretto." Through both gestures she reaches a condition of safe non-desirability ("inappetibilità"). Ultimately the fact that the reader does not witness Dirce's own disappearance is not only a clever narrative choice to close the story but leaves open the possibility that there may exist a (utopian) way (i.e., her dramatic and yet, as I have tried to argue, also "playful" gesture) to indeed break Kronos' pattern of appetite and consumption.

It may be that the kind of fictional preoccupations incarnated and expressed by Dirce and her sister may reflect quite closely similar, more realistic concerns in their author. Bontempelli's statements about the modern disappearance of the figure of the poet, on the weakened position of the intellectual in industrial, fascist society and, especially, his refusal to consider artistic work as merely some sort of merchandise inserted in the productive cycle (see note 285) suggest similar fears of untimely objectification and exploitation.

Like most artists of modernity, he had to deal with the anxiety of not wishing to be a commodity, of escaping an only for-profit, superficial consumption and yet, at the same time, of managing satisfactorily to deliver his artistic production and have it appropriately appreciated and "digested" by his audience. If this latter case could be an existential booster and contribute to sedating, at least temporarily, one's fears – just as the "successful" creation of Nora's baby, Fausto, momentarily saves her and her sister from desperation – the failure to fight and, possibly, avoid a merely profit-driven consumption seems, in the long run, only the premise for a premature and "in-fausto" (in the sense of both "inauspicious" and "fatal") sinking into oblivion.

As the powerful and memorable image of Dirce's begging subtly alludes one last time to the symbolic ties and relationships that the economic, the ludic, and the gastronomic have in *Gente nel tempo* and to the value that play has as a means to achieve a timeless "immortality", so does Bontempelli warn us also about the progressively marginal role that such a "kairotic" principle has in modern times. It seems only appropriate, therefore, that Bontempelli's goal and challenge in the next phase of his artistic career would be to keep shaping his "modernist critique of modernity" (Somigli 2004, 312) while continuing to play until the last of his own days his "magical" game of reconstruction and "recovery of the individual."[34]

34 As a final note, it may be appropriate to emphasize that: 1. "devi imparare a giocare" (one of the quotes placed at the beginning of this article), seals the conclusion of *Nembo* (1935), a suggestive theatrical *rappresentazione* by Bontempelli in which a creative game played by a group of children (or "artists") is threatened and then tragically interrupted by a mysterious, deadly dark cloud (possibly, fascism); 2. Bontempelli's 1930 novel *Vita e morte di Adria e dei suoi figli* opens precisely with the description of a game whose name "Liberi tutti!" is in direct contrast with Adria's subsequent behavior (she will enter a sort of self-imposed *clausura*) and her fear of the effects that a "kronos-like" ravaging time will have on her beauty. On the notion of *poeta ut puer ludens* in the latter novel see Bosetti's chapter dedicated to Bontempelli: "Émerveillement de l'enfant et étonnement poétique dans le 'réalisme magique' de Bontempelli" in Gilbert Bosetti, *L'enfant-dieu et le poète. Culte et poètiques de l'enfance dans le roman italien du XXe siècle* (Grenoble: ELLUG, 1997), 169–89, (184).

4.2 Milk and Blood: Nursing Capitalism in Pirandello's *La Balia*

The stunning conclusion of *Gente nel tempo* in which Dirce's character plays the role of beggar, and embodies a situation of progressive alienation from her surroundings, "esito estremo del processo di disgregazione della personalità [...] [and of the] effetti reificanti della società industriale di massa,"[35] exemplifies both Bontempelli's interest in addressing the effects of a modern consumeristic and profit-oriented society on individuals, and his familiarity with concerns and themes frequently associated with Luigi Pirandello's work.

Scholars often remind us that Pirandello's artistic work deals with the representation of a crisis in society related to the advent of industrial modernity and market capitalism at the beginning of the twentieth century. Romano Luperini, for example, writes that:

> la modernità appare a Pirandello come un'epoca in cui non sono più possibili né l'armonia tra uomo e cosmo del mondo classico, né le "correspondences" tra individuo e natura tipiche della cultura romantica e simbolista [...] È la fine della soggettività classica, ma anche di quella romantica. Non solo l'io appare svuotato di ogni fisionomia certa, ma l'interiorità stessa viene smantellata.[36]

More recently, Wladimir Krysinski adds that "Pirandello's work narrativizes, theatralizes and discourses upon the shattering of the bourgeois individual in Adorno's sense of the term (as a *Zerfall*, a fall and a division)," while Robert Dombroski, commenting on *Enrico IV*, writes about the "momentous transformation of individual subjectivity occurring under the impact of the capitalist market place; a fall into profound disorientation

35 Corrado Donati, "Il rapporto Pirandello-Bontempelli dalla trasgressione grottesca al richiamo del mito" in W. Geerts, F. Musarra, S. Vanvolsem, eds, *Luigi Pirandello: poetica e presenza*. Atti del convegno delle Università di Lovanio e Anversa, (13–16 maggio 1986), 404–6.

36 Romano Luperini, *L'allegoria del moderno* (Roma: Editori Riuniti, 1990), 223 and 226.

and insecurity that only madness can systematize."[37] In a modern (and modernist) world where, as Dombroski puts it, "the market is the organizing principle of culture and society" (134) and concepts such as self, subject, identity, and corporeality are often associated with ideas of instability, fragmentation, destruction and dissolution, the notion of consumption, because of its applicability to both the economic and physical spheres, may constitute a useful hermeneutical tool with which to offer new insights into modernism, in general, and into Pirandello's *oeuvre*, in particular. Rita Felski aptly observes that "apart from its economic meaning, consumption retains an association with exhaustion, waste, and destruction, signaling a process oriented toward the negation of matter and death" (Felski 76). In addition to these negative associations (which also inform its economic meaning), the term is linked to human alimentation and the physical body and the nutritional flows that (should) regulate its equilibrium. And the body, in turn, is the place where cultural values are inscribed. As Mary Douglas points out: "Certain cultural themes are expressed by rites of bodily manipulation. […] The rituals work upon the body politic through the symbolic medium of the physical body" (Douglas 128).

From this perspective, Italy's social-cultural situation in the post-First World War years, when the fascist regime was concerned with creating a "classless consumer who would inevitably identify his or her needs with a national economy […] tried to make something 'natural' and universal out of consumption […] pass[ing] itself off as a maternal entity" (Pinkus 83–5) and, in general, aimed at expanding and improving the nation's general health through a mixing of politics, biology, economics, and advertising is particularly interesting.[38] In an attempt to delineate the ideal body, certain foods themselves were "fascistized" (sugar, stimulants, grains). Pinkus, in her chapter "The Fascist Body as Producer and Consumer," discusses some

37 Wladimir Krysinski, "Pirandello in the Discursive Economies of Modernity and Postmodernism" in G.P. Biasin and Manuela Gieri, eds, *Luigi Pirandello. Contemporary Perspectives* (Toronto: University of Toronto Press, 1999), 218. Robert Dombroski, "Pirandellian Nakedness" (ibidem).

38 The representation of consumption and production in advertising is the object of Karen Pinkus' study.

aspects of the dietary transformations in this period and notes, for example, how in a particular advertising campaign "the cubes of sugar superimposed on the athlete [...] make a perfect, metonymic equation between the individual body and the body politic" (92) thus corroborating from another angle the centrality of the issue of alimentation in these years. Milk, and motherly milk in particular, certainly did not escape this process and that is why, as noted earlier, the fascist "nation-state was symbolically represented in the image of the breast-feeding mother" (Dixon Whitaker 95).

This focus should explain why Pirandello's "La balia," with its controversial representation of maternity and of the relationship between politics, lactation, and exploitation, is at the centre of my analysis.[39] In the pages that follow, I argue that the political-economic implications of the consumption motif in "La balia" metaphorically project the story well beyond the initial impression it gives of being a naturalistic-deterministic narration centered on the tragic vicissitudes of its female protagonist. Annicchia is a poor Sicilian peasant who, because of financial constraints, leaves behind her native land and newborn son (Luzziddu) in order to move to Rome and become the wet nurse for the weak baby (Leonida) of a wealthy Sicilian couple who now live there (Ersilia and Ennio Mori). Her actions substantially improve Leonida's health, but also tragically condemn her own offspring to death and, eventually, force her into prostitution.[40]

At first glance, an emphasis on alimentary consumption or an examination of the connection between physical and economic consumption may not seem the most productive way to approach "La balia." This short story

39 Luigi Pirandello, "La balia" (in "Il silenzio"), in *Novelle per un anno* ((Milan: Mondadori, 1990) (Vol I and II). Marco Bellocchio's decision to cinematically adapt Pirandello's short-story (*La balia*, 1998) is further proof of its status as a classic within the *Novelle*. The director, in the DVD interview, speaks of his desire to distance his cinematic adaptation from the "direzione zoliana, naturalistica veristica" which he identified in the original narration.

40 As Douglas Radcliff-Umstead observes: "Prostitution is the way out of economic straits [also] for the heroines of 'Volare' (1907) and 'Spunta un giorno'" (1927). Douglas Radcliff-Umstead, *The Mirror of Our Anguish: A Study of L. Pirandello's Narrative Writings* (Cranbury, NJ: Fairleigh Dickinson Univ. Press, 1978), 307, n. 14.

can certainly be interpreted in a standard, canonical fashion by focusing on those themes and motifs traditionally associated with Pirandello's artistic universe: form/formless, order/chaos, reason/folly, reality/illusion and so on.[41] And one might also object that Pirandello is not a modernist writer known for using an alimentary theme. But a quick look at a sampling of his other *novelle* shows that the alimentary/consumption motif present in "La balia" is not an isolated case.

If, for example, in "Un invito a tavola" a sumptuous meal allows a humorous look at Sicilian traditions and reveals both the excesses and the actual physiological dangers surrounding Sicily's culture of (returning) favors, in "Un'altra allodola" the combination of socio-economic and physical/alimentary consumption emerges more fully in the existential misery of two old friends who reunite after many years: "– Eh sí, – affermò Luca – è un fatto che, a me, il mangiare poco mi conferisce. Ma forse sono un po' troppo pallido questa sera, no? – Sei pallido, perché sei povero! – raffibbiò il Currao" (II, 345). The connection becomes even more convincing when Currao confesses that, in order to escape their poverty, his wife first betrayed him and then became a prostitute: "M'hai fatto le corna? T' hanno procacciato pane? Sí? E allora hai fatto benone! Danne un pezzetto anche a me!" (II, 345). Finally, "Come gemelle" shares literary ingredients with "La balia" in its narration of the vicissitudes of Marquis don Camillo Righi and the two baby girls he almost simultaneously produces, one with his wife and one with his lover. If, by dying in childbirth, Righi's wife represents a weakening and disappearing aristocracy (she is described as "magra, sgarbata, insoffribile"), his lover emblematizes the stronger, voracious rising class that will take its place in the socio-political order. As she proudly tells the Marquis: "Sono piú ricca di te [...] Cucio e mangio!" (I, 1193).

41 For example, one could see Annicchia's tragedy as the consequence of her folly, which is consisting of a rebellious departure from traditional habits, the will of her mother-in-law and biological laws: "E se sapesse come viviamo! Io, con le braccia legate da questa creaturina; lei vecchia di settant'anni! Volevo dare ad allevare il piccino e mettermi a servizio." (I, 430). Her abandonment of her son and Sicily emblematizes her desire for change, income, and a new life in the capital but, in the end, she is confined to another, even more stifling "form."

Correspondingly, the first baby is a "scheletrino" that even "tre balie" are unable to nurse, and the second one is the picture of good health. The conclusion of the *novella*, in which the lover offers to nurse Righi's weak but legitimate daughter together with her own, illustrates the lover's social climb, and her simultaneous transformation from seamstress who feeds off the wealth of the aristocracy, to someone who is their only source of subsistence and *raison d'être*; Righi himself "non poteva piú fare a meno di quella donna fervida e forte" (I, 1194). At the same time, this ending seems to allude (politically) to a near future when differences based on class and birth privileges will no longer exist: "Sembreranno proprio come gemelle, e non sapremo piú distinguere l'una dall'altra" (I, 1195).[42]

Similarly, the alimentary consumption in "La balia" alludes to socio-political and economic dynamics taking shape at the beginning of the twentieth century in post-Risorgimento, fascist Italy, and elsewhere in the Western world. At the poignant conclusion of this five-part *novella*, lawyer Ennio Mori's hypocritical and unrealistic words explicitly underscore the unfeasibility, if not the utter failure, of the socialist ideal, that is, the program which should have guaranteed to all people a fair share of goods and wealth, as well as basic, decent living conditions. Interestingly, the demise of this socialist-utopian dream according to which, "come diceva il Malon" (I, 453), all men are equals, is symbolically anticipated, both by the death of Annicchia's son who was starved so that his mother's milk could feed a socially superior baby and is therefore the victim of an indirect infanticide, and by the disappearance of Annicchia's old selves, both mother and wet nurse.

With remarkable prescience, "La balia" seems to address the inevitable, all-consuming rise of a proto-capitalist economy and its devastating effects on the progressively commodified, objectified bodies of the

42 Another novella dealing with "doubles," "I due compari" (also in *Tutt'e tre*), shows one more twist in the relationship between gestation and economics. In this story, the business partnership between Giglione and Butticè ends when the latter's wife dies while giving birth to their sixth child, thus unbalancing the condition of equality between them forever: "compenso era a entrambi la soddisfazione di sentirsi pari in tutto e l'uno degno dell'altro" (I, 1241).

weakest human beings. At the same time, Pirandello provides a critical commentary on a practice – that of breast-feeding – which was among the primary goals of fascist biological politics, and a powerful symbol of these ideas,[43] and subtly addresses the criminal, cannibalistic exploitation by the Italian North of its own South.[44] In this complex scenario, the appearance of Leonida's first tooth, together with the final "crazy" laughter of Ennio Mori's office clerk, "signor Ramicelli," may symbolize the advent of a new, aggressive and ravenous political and economic era. Ultimately, "La balia" (1923), which preceded "C'è qualcuno che ride" (1934), may be read as an "allegoria politica" and, as such, written at an even earlier point in time reveals that "anarchic nature [...] petit-bourgeois subversiveness of the generation which had established itself in the first two decades of the twentieth century," which Luperini specifically attributes to Pirandello in his later story.[45]

Let me now make a few additional remarks before taking a closer look at the text. First, I am aware that it is difficult, if not impossible, to provide a consistent political interpretation of Pirandello's work. Furthermore, I am equally conscious of the risks of suggesting a potentially limiting socio-political or ideological interpretation of a story that, at first glance, has as its focus an individual, rather than a collective, tragedy. As Gianfranco Vené reminds us: "la critica che Pirandello implicitamente porta contro le illusioni della società giolittiana non è [...] né di natura politica né sociologica

43 For a more detailed discussion of this topic, see Dixon Whitaker (95–7). It should also be mentioned that a partial autobiographical component may be traceable in the story, especially if one recalls the health problems of Pirandello's wife (and, thus, her difficulties raising their three children). See, for example, the representation of Ennio Mori's wife in Bellocchio's movie adaptation.

44 This is a theme, by the way, which he had already explored earlier in his most openly political novel, *I vecchi e i giovani*, 1913. In "La balia," however, given the Sicilian origin of most of the characters, there is also the more subtle and far-sighted intuition that the South actually largely contributes to its own exploitation/cannibalization, as if it were unaware of wasting its most vital nourishment.

45 Romano Luperini, "Laughter and Political Allegory in Pirandello" in Gianpaolo Biasin and Manuela Gieri, eds, *Luigi Pirandello. Contemporary Perspectives* (Toronto: University of Toronto Press, 1999), 107–24 (121).

[…] essa concerne la vita stessa dell'uomo al di lá delle contingenze storiche." (Vené 280). I do not wish to question the recognized critical truth that the individual subject rather than any social or political class is at the center of Pirandello's interest. But it also seems undeniable that the majority of the Sicilian writer's works target and denounce precisely the fundamental contradictions, paradoxes, and false sense of order belonging to the bourgeois, proto-capitalist social system, which he perceives as just another of the quintessentially tragic, repressive, and deadly forms which many human beings use to organize their short lives. As noted earlier, very few modernist artists possessed Pirandello's bitter and clear awareness of the "smarrimento del suo tempo e del suo precipitare verso il caos e l'abisso" (Vené 303) nor were they able to interpret so well the general sense of crisis of the times. Ultimately, then, as the same Vené concedes, for Pirandello man is political by nature, in the same way as he is naturally social, and, since the writer's problem is that of life itself, this is "un problema morale; e un problema sociale; e, nel significato più vasto, un problema politico" (Vené 359). To conclude these preliminary observations, some element of socio-political engagement should always be traceable when an artist repeatedly reveals to his audience the illusory and abstract nature of the law of society versus man's individual needs, feelings, and aspirations, or when he confronts readers with the representation of the unfairness of life and the tragedy of existence.

* * *

"La balia" opens in an upscale Sicilian household, where "la signora Manfroni," Ersilia's mother, theatrically comments on a letter sent by Ennio Mori, her Rome-based son-in-law, informing her of her daughter's recent childbirth. Her brief and progressively more irritated reaction to the epistle, alternating with the third-person comments of the narrator, allows the reader to gather immediately a large quantity of crucial information about the story's main characters and their reciprocal relationships.

We immediately learn about the ambivalent and resentful stance that socialist lawyer Ennio Mori has towards both his wife's family and the world. He first writes in detail about the dangers that Ersilia faced while giving birth, but he then hypocritically adds that it is not his intention to hurt or upset her relatives with this information. Rather, he just feels the need to complain about his wife's stubbornness and her habit of wearing "il busto troppo stretto, i tacchi delle scarpe troppo alti" (I, 425). The reader also finds out that Mori decides to hire a Sicilian (rather than a local, Roman) wet nurse. Although he claims that social-economic considerations and political principles are at the root of his decision, it is evident from his words that his fundamental stinginess is the main reason for his selection of Annicchia over someone from the capital city.

The allusions to Ersilia's obstinacy, her fashion preferences during pregnancy and, later on in the letter, to the contents of her dowry inform us of the wealth of her family, the nature of her marriage, her vanity and obsession with appearances, and her self-centered, unhealthy attitude towards maternity. Finally, Signora Manfroni's energetic verbal response to the letter reveals the very low esteem in which she holds Ennio's superficial socialist ideals ("Asino! Asino! Asino!"), her concern (similar to her daughter's) for always cutting a "bella figura" and, last but not least, her haughty attitude towards that lower, ignorant, and dirty social class "da lavare a sei e a sette acque" (I, 425) from which the wet nurse would unavoidably come.[46]

Once the narrator concludes his description of Signora Manfroni's outburst, it does not seem accidental that the first mention of the word "balia" is immediately followed by a reference to eating, as Signora Manfroni's bossy husband appears in front of her shouting: "Ohè! Non si mangia oggi? Perché la tavola non è ancora apparecchiata?" (I, 426). Two more allusions to food and nutrition in the next few lines confirm that this connection is not coincidental. The alimentary, political, and economic spheres become even more closely intertwined when Signor Manfroni, after reading the letter himself, determines who would be an appropriate wet nurse for his daughter, "La moglie di Titta Marullo" (I, 426). He then begins to voice opinionated judgments about the social and economic conditions of the

46 I observe that Mary Douglas' words on the essence of "the unclean" have a peculiar Pirandellian resonance: "Reflection on dirt involves reflection on the relation of order to disorder, being to non-being, form to formlessness, life to death." Qtd. in Korsmeyer (181–2).

times. Signor Manfroni's dialogue with his wife, in fact, reveals that his wealth derives from ownership of a bakery – from providing, but also partially controlling and regulating, the main source of nourishment for everyone in his town, including "tutta quella canaglia" who, by espousing socialist-revolutionary ideas, is very different from him both in terms of class and political opinion. Signor Manfroni further clarifies his socio-economic and political position when he explains to his wife the well-calculated logic behind both his firing (and implicit starving, since he stops paying/feeding him) of Titta Marullo ("lo scacciai dal panificio, per le sue idee rivoluzionarie"), and his subsequent acceptance of his daughter's marriage to Ennio (also a "revolutionary socialist" – at least "on paper"):

> E perché gli ho dato mia figlia, io? Prima di tutto perché Ennio è un ottimo giovine; poi, sissignora, perché socialista! sissignora! E mi è convenuto! e mi ha fatto gioco! Sai dirmi perché sono tanto rispettato io, da tutta quella canaglia a cui do da mangiare? […] Parlavamo di Titta Marullo […] Ora io, ricco, ma con qui dentro qualcosa che batte […] prendo sua moglie, la ficco in un vagone di terza classe e la spedisco a Roma, balia del mio nipotino! (I, 427).

Signor Manfroni's self-description is that of a hard-working and compassionate individual. However, the reader also easily infers that he is calculating, cynical, and self-serving. He is the perfect representative of that rising class of proto-capitalists who, unlike the aristocracy of the past, are building their fortunes on the skillful cultivation of their entrepreneurial, commercial, and political abilities.[47]

At this point, then, one might suggest that the initial section of the *novella* both introduces and implicitly confronts two radically different kinds of producers (or "feeders") and, by extension, two opposing economic models.[48] The first, represented by Signor Manfroni and, in general, by his extended family, could be defined as proto-capitalist. It feeds people, either with bread (in exchange for money), or money (in exchange for their

47 In this sense, Signor Manfroni could be an ideal literary ancestor of the character Don Calogero Sedara in G. Tomasi da Lampedusa's *Il gattopardo*.

48 Maud Ellmann observes that "eating comes to represent the prototype of all transactions with the other, and food the prototype of every object of exchange" (53).

services), but a profit, either in economic or existential terms, is always the goal. If there is no gain in view, or if such profit could be endangered by some competitive ideology, this model has the power to interrupt such literal and symbolic nutrition, thus revealing its true voracious nature behind the façade of "longanimità" and putting those whom it previously sustained into serious trouble (i.e., Titta Marullo, who ends up in jail or, in the end, Annicchia, when she is fired because she is no longer able to breast-feed).

Embodied by the "balia," the second economic model does not on principle exclude the possibility of an existential or monetary profit in order to improve present living conditions. Annicchia's desire to be hired as a wet nurse, after all, initially derives from her misunderstanding of the role Signor Manfroni might have in freeing her imprisoned husband, and her nursing contract provides a monthly remuneration she can send directly to her mother-in-law, "la vecchia Marullo." However, not only does the latter, as the more extreme representative of this second, anti-capitalist position, reject the money and return it to the sender ("piuttosto [...] sarebbe morta di fame o sarebbe andata a mendicare di porta in porta un tozzo di pane" [I, 442]) but, after her own son dies and she has stopped producing milk, the same Annicchia is no longer interested in receiving any of her rightfully due earnings: "[Ersilia] teneva a disposizione di lei il danaro che la suocera aveva rifiutato; ma Annicchia le rispose: 'E che vuole che me ne faccia più, oramai? Non ho più che questo qua, ora!'" (I, 450).[49] Annicchia's remark confirms that the nutritional, renewable economy she incarnates is not founded upon the acquisition of mere profit, convenience, or rationally calculated accumulation, but rather primarily on a truly compassionate, emotion-based disposition. Given that she represents this type of economy, she may also represent a truly proto-socialist economic model typical of an older, more convivial culture, in which resources (in this instance, milk as a nutritional capital), rather than being monopolized,

49 With the expression "questo qua" Annicchia refers to Ersilia's son (Leonida). However, given the conclusion of the story, the expression could also refer to her productive bosom, which, together with her body, is the only capital she has left.

exploited, and consumed, are instead produced, shared, and given away for free in times of surplus.

It is this latter sort of disposition and economic model that emerges, for example, when Annicchia, still undecided about her departure because of her son, discovers that the women in her neighborhood are ready to step in and take care of her baby:

> Una, anzi, ch'era venuta col suo figliolo in braccio, attaccato a un'enorme poppa: – Qua! qua! guardate, – si mise a gridare, – ho latte per due! Me lo piglio io, il bambino ... Qua, guardate! E, cavando il capezzolo di bocca al poppante, sollevando con una mano la mammella, fece sprizzare il latte in faccia alle comari del vicinato che, ridendo e riparandosi con le braccia, si scostarono addossandosi l'una all'altra (I, 431).

Unlike Signor Manfroni's bread, not only is milk supplied at no cost, as the product of a completely disinterested gesture, but it can even be joyfully wasted, thus acquiring a playful, cultural, and social-bonding function that surpasses its primarily alimentary one. The encounter, or better, the contrast between these two antipodal nutritional-economic positions and the effects, in terms of consumption, the clash has on individuals lies, I believe, at the narrative core of "La balia."

After what has been observed so far, it should not be surprising that Pirandello chose to situate the first dramatic representation of this clash, the meeting of Signor Manfroni with Annicchia, at a particularly emblematic moment in the plot: "Figurarsi che, mentre [il signor Manfroni] era ancora a tavola con la moglie, vide irrompere nella stanza Annicchia, la moglie di Titta, piangente di gioja, con un bambinello in braccio di circa due mesi" (I, 428). In other words, it would be difficult to imagine a more ideal moment in which to emphasize this confrontation than when "il signor Manfroni" and his wife are "a tavola," engaged in eating, which, on the one hand, epitomizes the fact that their class views reality in terms of consumption and, on the other, symbolically prefigures the similar behavior that their relatives in Rome will soon display towards Annicchia by ultimately reducing her to a sort of variously exploitable, metaphorically "edible" commodity.

The physical description of Ennio Mori as he waits for Annicchia's arrival at the train station warrants notice, as it seems to allude precisely to the pathological and hard-to-satisfy appetite that characterizes him:

"Piccolo di statura, magrissimo, con le spalle in capo [...] la faccetta ossuta, dalla tinta itterica, invasa e quasi oppressa da una barba nera troppo cresciuta [...] le tasche del pastrano e della giacca piene di giornali" (I, 431). His thinness and hunched posture strongly recall the image of a vulture waiting for its prey, and his nervousness and impatience, combined with his need to collect and then read all of the main Italian newspapers from cover to cover each day, reveal that an attitude fundamentally based on consumption affects his existence and approach to the world:

> Seguitava a leggere [...] e, ogni giorno, non si dichiarava soddisfatto, Se non aveva scorso da capo a fondo tutti i fogli più in vista di Roma e di Milano, di Napoli, di Torino, di Firenze, di cui aveva sempre così piene le tasche. – Medicina, – soleva dire. – Mi muovono la bile. Troppo, però! Eh, glielo aveva detto anche il medico (I, 432–3).

Even when Ennio, in order to pass the time, complains to himself about the unhappy condition of his marriage to Ersilia, he concludes his reflection with an economic consideration that betrays his far-from-socialist understanding of life: "Ah, parola d'onore, aveva fatto un bel guadagno a sposarla!" (I, 432). Clearly, he views marriage as an occasion for selfish accumulation and profit.

Towards the end of the story, moreover, the revelation that his seemingly heartfelt and humanitarian rhetorical attempt to convince Ersilia to keep Annicchia in their service after her personal loss is nothing but the spoken version of a lecture on socialism he had just written, proves to be one more instance of his tendency to devour, digest, and then spit out (for his own advantage and socio-political gain) even the most tragic of experiences:[50]

> Riflettendo intorno al triste caso di quel bambino morto laggiù in Sicilia, aveva pensato a un passo dell'opera del Malon *Le socialisme integral*; e, invece di farsene un rimorso, s'era proposto di farne argomento d'una conferenza che avrebbe tenuto al Circolo Socialista fra qualche giorno (I, 452).

50 As Radcliff-Umstead puts it, Ennio "delivers eloquent addresses on equality of opportunity but remains utterly indifferent to the wet nurse's plight" (Radcliff-Umstead 94).

Rather than presenting this sad episode as an opportunity for self-criticism or reconsideration, the narrator clearly points out that Ennio literally feeds on it, first to stir his imagination, and second, to sell his ideas better, and increase his personal and professional capital among his fellow party members at the "Circolo." If it is true that Ennio personifies the indifference and the ambivalences of his socio-political class and, in turn, the bulimic side of a dominant economic system whose basic nature is to figuratively gorge on and exploit everything and everyone to its own advantage, his wife Ersilia seems to complete the picture by embodying the equally pathological and destructive anorexic side of such a system.

Using (again) the term "anorexic," this time in the context of the scientific/literary observations made by Susan Skubal and Inge Lanslots, I do not mean only that Ersilia rejects her role as a mother and refuses both maternity and fertility.[51] The refusal of fertility derives from her lack of milk and *joie de vivre* and its cause – whether an inability or unwillingness to nurse – does not seem relevant. I also wish to suggest that her extreme spending habits, her concern for appearances, her obsession with rules and order, her compulsion for cleanliness, her longing for life combined with the corrosive jealousy eating away at her soul and, last but not least, her fundamental emptiness, contribute to delineate the picture of an unhealthy (if not actually "starving") woman and, in turn, of a whole system of economic power. That is to say a person or system, who or which, despite its wealth, shows substantial weaknesses and lacks something crucial, which in turn results in an inability to provide any sort of truly rewarding, literal, or symbolic nourishment to others as well as to her/itself.

In this light, Ersilia's repeated allusions to Annicchia's personal level of hygiene ("Andrai a lavarti […] – sei tutta affumicata […] Il fumo della ferrovia e le lagrime versate alla stazione le avevano insudiciato il volto.

51 See Susan Skubal's assertion that "the anorexic can't be mother" in: Susan Skubal, *Word of Mouth. Food and Fiction after Freud* (New York: Routledge, 2002), 74, and Inge Lanslots' similar remark that "l'altra conseguenza dell'anoressia è il rifiuto della fertilità e consecutivamente della maternità," Inge Lanslots, "Un assaggio della narrativa italiana contemporanea: il sapore di chiusura" in: Bart Van den Bossche and others, *Soavi Sapori*, 367–82 (377).

[...] – Basta, a lavarti, ora! [...] non toccare il velo con le mani sporche." [I, 436–7]), or her request for a doctor to examine the quality of the nurse's milk ("Il medico! Voglio subito il medico! – insistette" [I, 438]) acquire an extra layer of meaning.[52] These allusions, in fact, confirm her fear of contamination and her personal need for order, but also betray her irrational resistance to the idea that a thoroughly natural, simple, safe and "non-medicalized" (i.e., "dirty," primal and "organic") relationship may actually exist between (food) producers and consumers. Pushing this line of interpretation a step further, one may perhaps even suggest that those who are different, who belong to the other and are "unclean" (with all the socio-political, class connotations these words may have) will never be included in any sort of contract, be it nutritional (such as that of "baliatico") or social-economic.[53] The following passage, in which the narrator emphasizes how chronically sad and unsatisfied Ersilia is, further illustrates my previous remarks:

> Quantunque non ricevesse mai visite e di rado uscisse di casa, pure spendeva enormemente per gli abiti, dei quali alla fine restava sempre scontenta, come di tutto e di se stessa. Si sentiva, ed era forse davvero infelice; ma di questa sua infelicità incolpava gli altri, anziché la propria indole scontrosa, l'aspro carattere, la mancanza di ogni garbo. Era convinta che se si fosse imbattuta un altr'uomo che l'avesse amata e compresa, non avrebbe sentito tutto quel vuoto che sentiva dentro e attorno a sé. Ora le era venuto in uggia finanche il bambino, perché questi dimostrava di voler piú bene alla balia che a lei [...] [il marito], per quanto dicesse o facesse, non sarebbe riuscito a ispirarle, a comunicarle quell'affetto per la vita, di cui ella sentiva il desiderio smanioso, ma del quale nello stesso tempo la riteneva incapace. Se l'aspettava dagli altri, la vita, senza intendere che ciascuno deve farsela da sé (I, 444).

It is, indeed, tempting to view Ersilia as the personification of capitalism's weaker and sick side, its shortsightedness and paradoxically self-consuming,

52 Tangentially, since Ersilia requests a "medical intervention in maternity" (Dixon-Whitaker, 96) she conveys a rather fascist position on the matter.

53 Mary Douglas' reflections on "the essence of the unclean" as "that which violates its natural category by becoming its opposite" (qtd. in Korsmeyer, 181), seem particularly relevant to the context of a narrative in which a physically dirty wet nurse ends up becoming a morally unclean prostitute.

wasteful attitude. When the wet nurse arrives at her home, offering nutrition for her baby and the opportunity to rekindle an old friendship,[54] Ersilia, despite her "desiderio smanioso" and belonging to a class of quintessential consumers, fails to realize the long-term existential gain obtainable through keeping Annicchia. Ersilia thus fails to capitalize on the life-changing opportunity she is given and, instead, lets it literally waste away. Additionally, her joyless and unsatisfactory squandering of money to purchase dresses for herself stands in stark contrast with the gesture of the poor village woman who joyfully shared her milk for the benefit and satisfaction of many. In contrast to the generous conviviality of the poor woman's act, Ersilia's irresponsible, self-rewarding shopping, together with her ineffective attempt at disguising her sadness with clothes indicates the sterility and inefficacy of the voraciously consumerist – and yet existentially vacuous and lost – socio-economic class she represents. If, Pirandello seems to imply, the socialist ideal exists only at the rhetorical level instead of in reality, its alleged capitalist alternative is destined to fail all the more so. Despite (or, rather, because of) its all-consuming appetite, the consumptive capitalist model is just a new "forma," similarly failure-oriented, which can only damage, alienate and oppress the individual.

Capitalism's consumptive dynamics victimize certain individuals more than others and, clearly, Annicchia and her family belong to a category of unfortunates. However, Annicchia herself, besides being a victim of such dynamics, could also be considered the most telling example of the wide-reaching effects of the cannibalistic economic logic that, by reducing bodies to objects of exchange and consumption, regulates most human relationships in "La balia." The answer Annicchia gives to Ersilia's question about her son's age is, in this sense, emblematic:

> Quanto tempo ha il tuo figliuolo? – L'ho comprato, – rispose Annicchia, sforzandosi a guardarlo in volto, – che saranno due mesi.
>
> L'hai comprato? che dici? – Come debbo dire? – Ma fatto, figliuola mia, fatto ... I figliuoli si fanno ... si fanno ... Che c'è di male? (I, 439).

54 Annicchia "era cresciuta in casa ed era stata compagna di giuoco d'Ersilia, tant'anni" (I, 429).

Annicchia's use of the verb "comprare" to indicate the coming into existence of her son and thus his age, is relatively expected, despite Ersilia's condescending and surprised reply: the past participle "comprato" is the standardized version of a Southern Italian dialect expression ("accattato") still common among lower, uneducated classes. This expression is a euphemism that attempts to dilute, and thereby eliminate, the perceived shame connected to pregnancy and childbirth accruing to women in a Catholic country.[55] But in the wider context of a story in which the production and consumption of food, bodies, and economic-political models are closely intertwined, that same expression, especially if read together with Titta Marullo's exclamation when he appears at the Mori's door to furiously inform Annicchia about the death of their own son ("E tu ora, che vuoi fare? Me lo paghi o vuoi darmi il tuo" [i.e., Ersilia's]), acquires an eerie additional meaning. The commercial language of purchases and acquisitions used by Annicchia and her husband to refer to Luzziddu, in fact, reflects a popular, traditional understanding of children (especially strong, male children) as an investment which would later pay off, contributing to the family's future survival and the improvement of its work-force and wealth. In addition, however, their use of the language of commerce underlines the fact that both characters, despite their condition and status, are already unknowing elements of a society that considers bodies as objects/goods to own, exchange, and exploit. A society, that is, affected by the "disease of consuming other creatures' lives and possessions."[56]

55 I am told by a Sicilian friend that "rientrare in santo," for example, was another expression that indicated the notion that women, after being tainted by pregnancy and the shameful act of giving birth, should return to a (Catholic) ideal state, that is, to a pure, chaste, and immaculate condition.

56 Pierpaolo Antonello, "Il pasto nudo. Antropofagie postmoderne e trasformazioni del corpo," *Contemporanea. Rivista di studi sulla letteratura e sulla comunicazione*, 2 (2004), 122–43 (140). Annicchia's reflections and reasoning in terms of gains and losses after the death of her son could be another example of this condition: "'Sì, ma deve pure comprendere' pensava Annicchia, 'che il suo [Ersilia's] figliuolo appartiene, ora, anche a me: che se ella ci ha messo la pena di farlo, io ci ho rimesso il figlio per lui: e ora non mi resta più altro'" (I, 451).

It is worth remembering that Titta's announcement about Luzziddu's mysterious death (the cause of which is significantly left vague) comes after the narrator's mention of the appearance of *Nònida*'s [Leonida's] "primo dentuccio." This event suggests that, after feeding on Annicchia's milk, Leonida's appetite and that of the voracious social-economic class he symbolically embodies is now ready to switch to a heartier, carnivorous diet, one in which Luzziddu will be among the first to be consumed.[57] As anticipated, Annicchia herself will ultimately be the victim of a similarly predatory, all-consuming appetite when she is "swallowed" by the older, sharper teeth and lecherous "ugola" of "signor Felicissimo Ramicelli." Since his first appearance in the story it seems evident that the old Ramicelli, with his "lezii da scimmia" and odd blond hair, represents the latest and crudest tragicomic manifestation of opportunism and calculation: "rizzandosi le punte dei baffetti ritinti, andava a seder su la panca della sala d'ingresso, con la speranza che vi capitasse Annicchia, la bella balietta siciliana" (I, 445). For him, Annicchia represents an appetizing morsel to be devoured, someone who can be bought and who, because of her wet nurse credentials, should also be able to satisfy a very different kind of appetite: "Quell'Annicchia, per esempio, che bocconcino! Ogni qualvolta la vedeva, si sentiva toccar l'ugola" (I, 445).

At the end of the story, Annicchia's process of progressive commodification, her transformation from a "fresca e rosea" (I, 438) producer of nutritious breastmilk into a sort of stale food ("andat[o] a male", [I, 445, 453]) suitable only for adult consumption; from nurse to prostitute (that is, a figure who fully belongs to the "meccanismi di scambio, propri della logica di mercato", [Antonello 135]), becomes symbolic of the relentless system of replacement of a social-economic reality based on equality among human beings with one which, instead, functions according to the ravenous dynamics of domination and exploitation.

In this sense, there could be no more appropriate conclusion to "La balia" (and to this last chapter) than the diabolic laughter with which

57 "Ah, se almeno avesse potuto sapere con certezza come, perché fosse morto il suo bambino, se per mancanza di nutrimento o per qualche male non curato" (I, 451).

Ramicelli reacts to Ennio Mori's eloquent but indifferent political speech on socialist equality: "– Signor Ramicelli! – Avvocato! – Che ha? È impazzito? Perché ride così?" (I, 454). One does not have to dwell too much on the significance of laughter in Pirandello: as his famous reflections on the categories of the comic and humor remind us, laughter usually derives from the perception of something incongruous ("L'avvertimento del contrario")[58] and it is "the expression of a social censorship."[59] In general, therefore, laughter has been viewed as a "critical tool through which the modern societies that resulted from the technological revolution brought about by capitalism could be criticized" (Pagano 46).

In line with the observations above, then, Ramicelli laughs first and foremost because he perceives the discrepancy between Ennio Mori's real behavior and attitudes and his compassionate-sounding (but actually empty) words on socialist equality and social justice. In other words, his laughter derives from his perception of an incongruity, from the realization that a rhetorical "form" has actually been imposed on "life." Ennio Mori's words signal the opposite of what his life as a consumer/exploiter is and, therefore, of what he would say if he were more honest with himself and a less hypocritical politician. One could also add that Ramicelli's laughter betrays his happy anticipation of eventually "tasting" Annicchia himself, after having hired her as his servant/mistress, and simultaneously sends a knowing wink to Ennio Mori, as a previous, fellow profiteer of her.

Ultimately, however, it is hard to overlook something different and more disturbing in Ramicelli's laughter, especially if one recalls Luperini's analysis of "C'è qualcuno che ride" as "a political allegory" with "sinister and perturbing tones" and an example of "the directions taken by Pirandello in his overcoming of the poetics of humor" (Luperini 120–1). As in "C'è' qualcuno che ride," the conclusion of "La balia" does not seem to belong to the category of comic or humorous narration. As a consequence, even

58 Luigi Pirandello, *Saggi, poesie, scritti vari*, ed., Manlio Lo Vecchio Musti (Milan: Mondadori, 1993), 127.

59 Tullio Pagano, "Modernisms: From Bergson's Laughter to Pirandello's Humor," *The Italianist*, 17 (1997), 44–59 (53).

more than representing laughter as a "tool" with which to criticize the outcomes of capitalism, the final close-up image of Ramicelli's wide open, toothy, all-devouring mouth seems to be the frightening confirmation and the most telling sign that consumptive capitalist dynamics have become so widespread and overwhelming that identifying and/or criticizing them is virtually impossible. What "La balia" ultimately tells us, therefore, is that the soft and toothless baby's mouth nursed by Annicchia has now grown into its adult version, blood-thirsty and ready to consume anyone who comes within its range.

Conclusion

The analysis of various literary representations of alimentary discourses – from their explicit manifestation in *Le Roi Bombance* and *The Futurist Cookbook* to their more nuanced appearances in the other considered works by Marinetti, Palazzeschi, Bontempelli, Masino and Pirandello – reveals their pervasiveness and relevance within Italian modernism and, by extension, modern Italian literature's participation and contribution to a broader discourse centered on the body, its dynamics, its potential and its failures in rapidly changing cultural circumstances. As a consequence, discussing the aesthetic and metaphorical function of food in these texts has also meant to reflect on some of the salient features of modernism and modernity, and their often contradictory impulses and concerns.

Among these, one may recall the potential and limits of the body (biological and social, private and public) and of aesthetic expression in a progressively more commodified and technology-oriented world; the controversial relationship with the past and (futurism's) attempt at metabolizing and processing it; the changing perception of the role of the intellectual/artist in the political and cultural climate generated by totalitarian and colonial regimes; and the ambivalent perception of the feminine body as both producer and consumer within the emergence of mass culture and capitalist economies. In several of the texts under discussion we see, on the one hand, an artistic tendency to bulimically gorge on, and then creatively "regurgitate" a shifting social and material reality in the utopian hope of getting a better grip on it and on one's own self and, on the other, a simultaneous and even more frequent, antipodal desire to "resist accumulation" (Armstrong 56), to withdraw from "History" and "anorexically" starve according to a self-negating "economy of sacrifice [...] an implacable aesthetic that demands the decreation of the flesh" (Ellmann 59), in order to paradoxically achieve a similar objective of artistic and subjective self-fulfillment.[1]

While it may be redundant, at this point, to stress the degree to which the alimentary trope pervades modernist literary aesthetic, it may be useful to reiterate that Italy's literary experience of modernity and the corresponding sociohistorical process of industrialization, which began in the years following its political and cultural "Unification" as a nation, was different and delayed in comparison to that of several other Western countries, and that this situation too, in turn, may also partially explain the resistance to "modernism" as a cultural and critical category in Italy. The combination of Italy's geographical diversity and sociopolitical fragmentation, its lasting lack of unity (and, thus, of any real sense of the State), the traditional hostility of the Catholic Church towards progressive philosophies and new ideologies, fascism's role in "weaken[ing] the ties linking Italian modernism to its European counterparts," and the influential presence of figures such as Croce and Carducci who "could [not] conceive literary modernity outside the context of Classicism" (Somigli and Moroni 2004, 13–17), are some of the factors that contributed to the peculiarity of the Italian situation. As I have argued, however, it seems undeniable that, despite Italy's slowness in catching on to the modernization process, there are some advantages in adopting the term "modernism" over the various, other "isms" commonly used to define the period.

Within the focus of this exploration, perhaps the most evident, practical benefit of viewing a selection of early twentieth-century Italian writers under the modernist label is that it has allowed me to literally bring different "foods to the table." With this expression, I wish to emphasize not only the previously mentioned interest of modernism in issues of corporality, the "regulation and clarification [...] of the body" and its relation with writing

1 In Ellmann's words: "It is by ingesting the external world that the subject establishes his body as his own, distinguishing its inside from its outside. If the subject is founded in gustation, though, this also means that his identity is constantly in jeopardy, because his need to incorporate the outside world exposes his fundamental incompleteness [...] The catch is that the very need to eat reveals the 'nothing' at the core of subjectivity" (30).

and textuality (Armstrong 5–7), but also the "opening-up" of the possibility of approaching certain authors and texts from a critical perspective (i.e., "gastro-criticism") in which the body, the food that goes into (and out of) it and, in general, various kinds of alimentary references may be a valid instrument for interpretation. In other words, while the work of several modernist writers belonging to other nationalities and literary traditions has been more or less "regularly" read from this critical angle – it should suffice to check the number of books and articles devoted to the meaning and implication of alimentary images in the texts of Kafka, Yeats, Joyce, Stein, Beckett, Woolf, Hemingway, Dostoevsky – the work of Italian writers who were active at more or less the same time as these well-known masters of modernity has (with a few noted exceptions) rarely been addressed in such a way. Showing the potential and critical fruitfulness of a similar kind of analysis in the modernist Italian context and, at the same time, considering for once the combination of the terms "Italy" and "food" as capable of generating acceptable literary and cultural research rather than the usual plethora of stereotypes and superficialities one is frequently exposed to, have obviously been among the main goals of this study.[2]

It goes without saying that this work, in its limits and selectivity, leaves room for further investigation and expansion on the topic. In the rest of this final chapter, I would like to suggest two of the potential directions such expansion could take.

First, and most expectedly, one could mention additional Italian modernist writers who may themselves be approached from a more explicit

2 To clarify what I mean when I say "stereotypes" and "superficialities," a personal anecdote may help. On a recent trip from Italy back to the U.S.A. (still as a green-card holder/non-immigrant alien) I was "semi-jokingly" asked by an officer at customs if, being Italian, I had not brought back and packed some sausages and salami in my suitcase (I had not). I imagine that a French person would be asked about cheese; a Chinese person about rice, and so on. I realize that this may not seem important but this kind of comment indirectly contributes to what is ultimately an unhelpful cultural pigeon-holing achieved "via food" (one which, in my particular instance, evokes the whole history of Italian immigration to the U.S. and subsequent dynamics affecting race and power relations).

alimentary angle. The names of Carlo Emilio Gadda and his "macaronic" texts, Italo Svevo and his own version of a "Kafkaesque" "rhetoric of renunciation,"[3] Annie Vivanti and her phagocytizing vision of family dynamics in *The Devourers* (1910) have surfaced before, both here and in other critical contexts.[4] Two additional (and extremely different) modernist Italian authors who come to mind are then Alberto Savinio and Federigo Tozzi. The former's short story "Il Signor Muster," for example, in which the character's body progressively falls to pieces, could be read as an extreme case of "anorexic" self-effacement (not to mention as an archetype of post-human dissolution). Tozzi's fictions, in turn, are often characterized by what could be called a "poetics of disgust" in which the biological and alimentary dimension plays a relevant role (I am thinking especially of certain passages about eating in *Tre croci* and the short story "Un'osteria"). Despite the obvious differences (formal, stylistic, genre-related) both writers often seem to be dramatizing potential glitches in the mediation between a subject and a world that often seems senseless and unable to provide any sustenance.

The second research direction, however, in its wealth of ramifications could ultimately be even more rewarding and further strengthen the relevance of a gastro-critical approach. Of course, this second critical stance would continue to interpret, through an even more encompassing alimentary key, works and cultural objects in which the semantic field of food is not necessarily explicit but is, instead, implicitly revealed and alluded to by the way they address and thematize the relationship between human beings and the world they live in, their "inside" vs. the "outside." Additionally, perhaps borrowing from some of modernism's own experimental techniques, this approach would not shy away from the possibility of individuating "jolts of connection" between different or apparently distant fields and of looking for simultaneous and multidirectional readings.

3 In one of his diary entries, he states that "Io a quest'ora [i.e., after the unflattering reviews of *Senilità* (1898)] e definitivamente ho eliminato dalla mia vita quella ridicola e dannosa cosa che si chiama letteratura," in Italo Svevo, *Racconti, saggi, pagine sparse*, in *Opera omnia*, Vol. III (Milan: Dall'Oglio, 1968), 818.

4 Biasin dedicates a chapter to Gadda in his *The Flavors of Modernity*.

One could thus continue by asking questions such as the following – could Slavoj Žižek's theory, according to which the "social real is incarnated in fantasy,"[5] also mean that, for instance, studying the ways in which alimentary discourses appear in literary texts may tell us something about the inner functioning, or metabolism, of such "social real"?[6] Is it possible that the recurring alimentary motifs emphasized in the previous chapters in connection with a modernist aesthetic (i.e., consumption, anthropophagy, anorexia, self-effacement, voraciousness) may also more broadly resonate with some (postmodern) contemporary representations of the body and the cultural and socio-political dynamics that alternatively "feed" it or "starve" it, both literally and metaphorically, physiologically and existentially?[7] Or, to put it differently, what may the various discursive practices explored so far tell us about the "rhetoric of consumption in our own time" (Frega 2)? Would it make sense to speculate that the analyses of the ways in which Marinetti's work incorporates an apparent "omnivorous" impulse towards reality (be it past, present or future) may reflect more global, contemporary attitudes towards the body? Is it appropriate to hypothesize that the pervasive "uncertainty and anxiety around food in modern society" (Morton 13) may be partially foreshadowed when Palazzeschi wishes to "eat the world" but also to simultaneously reduce the foundational value of reality (Cangiano 178); when Masino connects the dimensions of physiological and spiritual nutrition; Bontempelli deals with the devastating effects of a

5 Qtd. in Timothy Morton, ed., *Cultures of Taste/Theories of Appetite: Eating Romanticism* (New York: Palgrave Macmillan, 2004), 8.

6 Korsmeyer's answer to this query seems to be positive: "In the dark side of eating we discover horrifying possibilities that are the counterpoint to food as a civil and binding aspect of society. Indeed, certain aspects of the dark side are actually presupposed by the transformation of this activity into a civilizing and unifying practice" (189).

7 In his essay on Knut Hamsun's novel *Hunger* (1890) (in *The Art of Hunger*, 1992), Paul Auster notes that: "In the end, the art of hunger can be described as an existential art. It is a way of looking death in the face, and by death I mean death as we live it today: without God, without hope of salvation. Death as the abrupt and absurd end of life" (20).

persistent "devouring myth" on bodies and minds; or Pirandello addresses the fate of human beings in a time of neocapitalist commodification?

And, finally, what are the chances that recent cultural debates on interdisciplinary topics such as "sustainability," "human ecology," and "posthumanity" may be reverberating some of modernism's own food-related concerns and sensibilities as we have seen them emerging in these fictional fantasies?

While this conclusive chapter is not the context in which to provide in-depth answers to these disparate and heterogeneous questions, the considerations that follow seem to suggest that it is at least pertinent and potentially fruitful to ask them. For instance, the way in which, in the era of global capitalism, the "human body may be observed in terms of "passing through" […] a territory to be conquered [and] appears to be divided into zones of economic interest, much like that of a pig sectioned at the butcher shop"[8] is a first, powerful reminder of the resilience and range of alimentary metaphors, and of how Western society and culture continue to engage in more or less well-disguised post-colonial, "anthropophagic" dynamics of exploitation and subjugation.[9] Additionally, the disturbing realization that "in our post-human times" the body is perceived in terms of "flesh […] subject to increased control either in the laboratory or in the marketplace and is caught up in processes of modification that seek to master and profit from it […] relegated to the status of unwanted materiality that can be pushed aside"[10] brings the destiny of "human flesh" closer to the one so far reserved for animals and, as it implicitly questions the primacy of anthropocentrism, ideally resonates with futurism's own proposal to famously discard bodies "older than thirty" and futurism's vibrations

8 Dejan Atanackovic, in the brochure of *Non-corpi*, an exhibition presented by the Siena Art Institute and the Galleria Fuoricampo, 10–20 Dicembre 2011.

9 In his essay "'Eating Well,' or the Calculation of the Subject," Derrida uses the term "carno-phallogocentrism" to indicate the origin of Western subjectivity and of a culture that eats and, in turn, is being eaten by the other.

10 Zoe Detsi-Diamanti and Katerina Kitsi-Mitakou, in Katerina Kitsi-Mitakou and Effie Yiannopoulou, eds, *The Future of Flesh: A Cultural Survey of the Body* (New York: Palgrave Macmillan 2009), 4–5.

between the organic and the inorganic. Lastly, the proliferation of ecologic and environmental tendencies focussed on minimalist behaviors and the refusal of standard, "institutional" forms of sustenance (be they alimentary, spiritual, energetic) in order to experience a "cleaner," "leaner," and "purer" life, different from that of the "undifferentiated 'crowd'" (Heywood 63) may, in turn, resemble updated permutations of the "anorexic" patterns traced in Kafka, Palazzeschi and Masino.

If there are still doubts about this approach, what I would like to stress is that a gastro-critical approach to literature in general, and modernist literature in particular, may help not only to provide fresh textual interpretations, but also to reflect on how we live and how we create and, ultimately, to better recognize and adjust to (i.e., accept or resist) the infinite challenges that the body, either in its private or public dimension, will continue to face in the future. The inclusion of a specific Italian "modernist component" in this effort seems particularly appropriate, since it involves a place and culture in which, on the one hand, questions of subjective, national (and, even more, regional and local) identity often play out at the level of specific, tangible food choices and, on the other, is also often literally and metaphorically "consumed," be it visually, as the object of the voracious gaze of global tourists (who, in turn, feed the local economy),[11] or geographically and politically.

The conclusion of "Vecchie inglesi," a 1932 short story by Palazzeschi, may partially summarize, although in the irreverent, topsy turvy and allusive way typical of this writer, the sort of local vs. global, modernist vs. contemporary dynamics I have considered and may also provide a memorable culinary-sexual image with which to conclude this book. Palazzeschi's story ends with the representation of a global cultural changing of the guard and the perspective of a new economy of consumption that is staged in the microcosm of the streets of Florence. Here an anthropomorphized, consumed "old Europe" (England), "grassoccia e appetitosa per quanto

11 On the notion of "visual consumption," see John Urry, "The Consuming of Place" in Adam Jaworski and Annette Pritchard, eds, *Discourse, Communication and Tourism* (Clevedon-Buffalo: Channel View Publications, 2005), 19–27.

matura" looks at her beautiful, provoking young daughter (America), which is taking her place as the latest, primary object of desire for "gli sguardi avidi dei giovinotti" (Italy). As the old woman makes an effort to keep pace with her daughter and avoid becoming totally irrelevant, she sings and shouts "Io non ho più banane!" to which the young lady smilingly replies, "E io ne avrò finché vorró. Taratun ta ta ta teiro … Taratun ta ta ta ta …"

The possibility that, one day, bananas could run out also for the young woman, or that she too could eventually grow old and be outstaged by some other woman was probably difficult to conceive for Palazzeschi, at a time when Europe was still able to generally look at America as an attractive land of plentiful resources and unending opportunities. Of course, contemporary readers know better and, at the end of this culinary journey through Italian modernist texts, they should also hopefully be better able to savor such a metaphor in its full range of flavors.

Bibliography

Primary Sources

Adam, Paul, *Lettres de Malaisie* (Paris: G. Crès, 1922).

Bontempelli, Massimo, *L'avventura novecentista* (Firenze: Vallecchi, 1974).

——, *La fame* in *Teatro* Vol. II (1932–35) (Milano: Mondadori, 1947).

——, *Gente nel tempo* in *Racconti e romanzi.* Paola Masino, ed., Vol. II (Milano: Mondadori, 1961).

——, *Nembo* in *Teatro* Vol. II (Milano: Mondadori).

Fourier, Charles, *Le Nouveau Monde amoureux* (Paris: Slatkine, 1979).

——, *Le Nouveau Monde Industriel et Sociétaire* VI (Paris: Éditions Anthropos, 1966).

——, *Teoria dei quattro movimenti, Il nuovo mondo amoroso e altri scritti sul lavoro, l'educazione, l'architettura nella società di Armonia,* Italo Calvino, ed. (Torino: Einaudi, 1971).

——, *Théorie de l'Unité Universelle* IV (Paris: Éditions Anthropos, 1966).

——, *Théorie des Quatre Mouvements* (Paris: J. Jacques Pauvert Éditeur, 1967).

Kafka, Franz, *A Hunger Artist*, in *The Complete Stories*, N.N. Glatzer, ed. (New York: Schocken Books, 1971), 268–77.

——, *Diari 1910–23* (Milano: Mondadori, 1977).

Marinetti, F. Tommaso, *La grande Milano tradizionale e futurista* (Milano: Mondadori, 1969).

——, *Novelle colle labbra tinte*, in Domenico Cammarota, ed. (Firenze: Vallecchi, 2003).

——, *Patriotismo insetticida. Romanzo di avventure legislative* in Calendoli, Giovanni. *Teatro. F.T. Marinetti* Vol. I (Roma: Vito Bianco Editore 1960).

——, *Le Roi Bombance* (Paris: Société du Mercure de France, 1905).

——, *Re Baldoria.* Trans. by Decio Cinti (Milano: Fratelli Treves, 1920).

——, *Scatole d'amore in conserva* (Firenze: Vallecchi, 2002).

——, *Teoria e invenzione futurista* (*TIF*), Luciano De Maria, ed. (Milano: Mondadori, 1990).

—— and Fillìa, *La cucina futurista* (Milano: Longanesi, 1986).

Masino, Paola, "Fame" in *Colloquio di notte. Racconti.* Prefazione di Maria Rosa Cutrufelli. Trans. by Maria Vittoria Vittori, ed. (Palermo: La Luna, 1994).

——, *Nascita e morte della massaia* (Milano: La Tartaruga, 1982).
——, *Io, Massimo e gli altri. Autobiografia di una figlia del secolo*. Introduzione di M.V. Vittori (Milano: Rusconi, 1995).
Mussolini, Benito, *Scritti e discorsi*, Vol. II (Milano: Hoepli. 1934).
Palazzeschi, Aldo, *: riflessi* (Milano: SE, 1990).
——, *Il codice di Perelà* (Milano: SE, 1991).
——, "Il controdolore", "Equilibrio", "Varietà", respectively in *Lacerba*, anno II, n. 2, 15 gennaio 1914; anno II, n. 24, 1 dicembre 1914, and anno III, n. 1, 3 gennaio 1915.
——, *L'interrogatorio della contessa Maria*. Introduzione di Fabrizio Bagatti (Milan: Mondadori, 1988).
——, *La piramide. Scherzo di cattivo genere e fuor di luogo* (Firenze: Vallecchi, 1926).
——, *Stampe dell'800*. Enrico Ghidetti, ed. (Milano: Mondadori, 2003).
——, *Tutte le poesie* (Milano: Mondadori, 2002).
——, *Tutti i romanzi* (Milano: Mondadori, 2005).
Papini, Giovanni, "Chi sei?", in *Il tragico quotidiano* (Florence: Vallecchi, 1906).
——, *Lo specchio che fugge* (Milan: Mondadori, 1990).
Pirandello, Luigi, "La balia" in *Novelle per un anno* (Milan: Mondadori, 1996).
——, *Saggi, poesie, scritti vari*, Manlio Lo Vecchio Musti, ed. (Milan: Mondadori, 1993).
Settimelli, Emilio, *Donna allo spiedo* (Milano: Modernissima, 1921).
——and Corra, Bruno, *Prefazione a F.T. Marinetti 'Come si seducono le donne'* (Firenze: Vallecchi, 2003).
Svevo, Italo, *Opera omnia I. Epistolario* (Milano: Dall'Oglio, 1966).
Vivanti, Annie, *I divoratori* (Palermo: Sellerio, 2008).

Secondary Sources

Adamson, Walter, *Avant-garde Florence: From Modernism to Fascism* (Cambridge, MA, Harvard University Press, 1993).
Airoldi Namer, Fulvia, "La terra e la discesa: l'immaginario di Paola Masino", *Otto/Novecento* 24, 3 (2000), 161–86.
——, *Massimo Bontempelli* (Milano: Mursia, 1979).
Anderson, Mark, "Anorexia and Modernism, or How I Learned to Diet in All Directions" *Discourse: Berkeley Journal for Theoretical Studies in Media and Culture* 11, 1, Fall–Win. (1989), 28–41.

Andits, Eszter, "'Sore on the Nation's Body': Repression of Homosexuals under Italian Fascism," unpublished thesis (2010), 66 www.etd.ceu.hu/2010/andits_eszter.pdf. [Last accessed July 2012].

Antonello, Pierpaolo, "Il pasto nudo. Antropofagie postmoderne e trasformazioni del corpo", *Contemporanea. Rivista di studi sulla letteratura e sulla comunicazione*, 2 (2004), 122–43.

Armstrong, Tim, *Modernism, Technology and the Body. A Cultural Study* (Cambridge: Cambridge University Press, 1998).

Auster, Paul, *The Art of Hunger* (Los Angeles: Sun & Moon Press, 1992).

Baldacci, Luigi, "Interrogativi sulla Contessa Maria", *Paragone* n. 458 (1988), 3–16.

Barilli, Renato, *La linea Svevo – Pirandello* (Milano: Mursia, 1972).

Bataille, Georges, *Visions of Excess: Selected Writings, 1927–1939*. Edited and with an Introduction by Allan Stoeckl (Minneapolis: University of Minnesota Press, 1985).

Beecher, J. and Bienvenu, R., *The Utopian Vision of Charles Fourier* (Boston, MA: Beacon Press, 1971).

Benjamin, Walter, *The Arcades Project*, ed. Roy Tiedemann, trans. Howard Eiland and Kevin McLaughlin (Cambridge, MA: The Belknap Press, 1999).

Bennett, Jane, *Vibrant Matter. A Political Ecology of Things* (Durham and London: Duke University Press, 2010).

Berghaus, Günter, *Futurism and Politics: Between Anarchist Rebellion and Fascist Reaction 1909–1944* (Providence, RI: Berghahn Books, 1996).

Berman, Marshall, *All That Is Solid Melts Into Air. The Experience of Modernity* (New York: Penguin Books, 1982).

Biasin, Gian Paolo, *I sapori della modernità* (Bologna: Il Mulino, 1991).

Blelloch, Paola, "*From Trunk to Grave: the Hallucinated Story of a Housewife*", *Nemla Italian Studies*, 13–14 (1989–1990), 89–103.

Blum, Cinzia Sartini, *The Other Modernism. F.T. Marinetti's Futurist Fiction of Power* (Berkeley: University of California Press 1996).

Bohn, Willard, "The Poetics of Flight: Futurist 'Aeropoesia'", MLN, vol. 121, n. 1 (January 2006), 207–24.

Boothroyd, David, "Foucault's Alimentary Philosophy", *Man and World. An International Philosophical Review*, vol. 29, n. 4 (October 1996), 361–85.

Bosetti, Gilbert, "Émerveillement de l'enfant et étonnement poétique dans le 'réalisme magique' de Bontempelli" in Bosetti, G. *L'enfant-dieu et le poète. Culte et poètiques de l'enfance dans le roman italien du XXe siècle* (Grenoble, ELLUG, 1999), 169–89.

Bowler, Anne, "Italian Futurism and Fascism," *Theory and Society* 20.6 (December 1991).

Bradbury, Malcom and McFarlane, James, *Modernism. A Guide to European Literature 1890–1930*. London: Penguin, 1991.

Braidotti, Rosi, *Nomadic Subjects: Embodiment and Sexual Difference in Contemporary Feminist Theory* (New York: Columbia University Press, 1994).

Bridgwater, Patrick, *Kafka and Nietzsche* (Bonn: Bouvier, 1974).

Brose, Margaret, "Giuseppe Ungaretti's Sentimento del Tempo: Baroque Rome and the Experience of Time." *Pacific Coast Philology*, 21:1–2 (November 1986), 65–72.

Brown, Bill, "Thing Theory." Things. Ed. Bill Brown. Special issue of *Critical Inquiry* 28, 1 (2001), 1–22.

Brown, James W., "Alimentary Discourse in Nineteenth-Century Social Theory: Pierre Leroux, Etienne Cabet and Charles Fourier," *Dalhousie French Studies* 11 (1986), 72–96.

Caillois, Roger, *Man, Plays and Games*. Translated by Meyer Barash (New York, The Free Press, 1961).

Calinescu, Matei, *Five Faces of Modernity: Modernism, Avant-garde, Decadence, Kitsch, Postmodernism* (Durham, NC: Duke University Press, 1987).

Camporesi, Piero, *The Land of Hunger* (Cambridge: Polity Press, 1996).

Cangiano, Mimmo, *L'uno e il molteplice nel giovane Palazzeschi (1905–1915)* (Firenze: Società Editrice Fiorentina, 2011).

Cecchini, Carlo, *Avanguardia mito e ideologia: Massimo Bontempelli tra futurismo e fascismo* (Roma: Il Ventaglio, 1986).

Cirillo, Silvana, *Nei dintorni del surrealismo. Da Alvaro a Zavattini umoristi balordi e sognatori nella letteratura italiana del novecento* (Roma: Editori Riuniti, 2006).

Cirlot, J.E., *A Dictionary of Symbols* (New York: Dorset Press, 1971).

Comoy Fusaro, Edwige, "Réflexions sur le texte (pseudo-)littéraire chez Paolo Mantegazza: *Un giorno a Madera. Una pagina dell'igiene d'amore*," *Cahiers de narratologie* 14 (2008), <http://revel.unice.fr/cnarra/document.html?id=543>. [Last accessed October 2012].

Cooper, Allison, "Gender, Identity, and the Return to Order in the Early Works of Paola Masino" in Somigli, Luca and Moroni, Mario (eds). *Italian Modernism. Italian Culture between Decadentism and Avant-Garde* (Toronto, Buffalo, London: University of Toronto Press, 2004), 378–99.

Crary, Jonathan, *Techniques of the Observer: On Vision and Modernity in the Nineteenth Century* (Cambridge: MIT Press, 1992).

Curi, Fausto, "Una stilistica della materia." *Tra mimesi e metafora. Studi su Marinetti e il Futurismo* (Bologna: Pendragon, 1995), 47–79.

Curi, Franco, "Edipo, Empedocle e il saltimbanco" *Il Verri* 6 (1974), 58–140.

Dagonet, Francois, *Trois philosophies revisiteés: Saint Simon, Proudhon, Fourier* (Hildesheim: G. Olms, 1997).

D'Amico, Silvio, "The Italian Drama in the Past Five Years." English translation by J.H. Clancy. *Educational Theatre Journal*, vol. 2, n. 3 (October 1950), 208–17.

De Grazia, Victoria, "Nationalizing Women: The Competition between Fascist and Commercial Cultural Models in Mussolini's Italy", in *The Sex of Things; Gender and Consumption in Historical Perspective* (Berkeley and Los Angeles: University of California Press, 1996), 337–58.

Deleuze, Gilles and Guattari, Félix, *Kafka. Toward a Minor Literature*. Translation by D. Polan (Minneapolis-Oxford: University of Minnesota Press, 1986).

Delville, Michel, *Food, Poetry, and the Aesthetic of Consumption: Eating the Avant-Garde* (New York: Routledge, 2008).

De Maria, Luciano, *La nascita dell'avanguardia* (Venezia: Marsilio, 1986).

Detsi-Diamanti, Zoe, Kitsi-Mitakou, Katerina, Yiannopolou, Effie (eds), *The Future of the Flesh: A Cultural Survey of the Body* (New York: Palgrave Macmillan, 2009).

Di Fazio, Margherita, *La lettera e il romanzo. Esempi di comunicazione epistolare nella narrativa* (Roma: Nuova Arnica Editrice, 1996).

Direk, Zeynep, "Erotic Experience and Sexual Difference in Bataille." *Reading Bataille Now*, ed. Shannon Winnubst (Bloomington: Indiana University Press, 2007), 94–115.

Dixon, Whitaker, *Measuring Mamma's Milk: Fascism and the Medicalization of Maternity in Italy* (Ann Arbor: University of Michigan Press, 2000).

Donati, Corrado (ed.), *Massimo Bontempelli, scrittore e intellettuale: atti del convegno*, Trento, 18–20 aprile 1991 (Editori Riuniti, 1992).

Douglas, Mary, *Purity and Danger: An Analysis of the Concepts of Pollution and Taboo.* (London and New York: Routledge, 1991).

Duncan, J. Ann, *Les romans de Paul Adam: Du Symbolisme littéraire au symbolysme Cabalistique* (Berne, Francfort/M, Las Vegas: Peter Lang, 1977).

Eliade, Mircea, *Patterns in Comparative Religion*, trans. by R. Sheed (Lincoln and London, University of Nebraska Press, 1996).

Ellmann, Maud, *The Hunger Artists. Starving, Writing, and Imprisonment* (Cambridge, MA: Harvard University Press, 1993).

Eruli, Brunella, "Da un re all'altro: Jarry e Marinetti", *Franco-Italica: Serie Storico-Letteraria* 1 (1992), 145–60.

——, "Preistoria francese del futurismo", *Rivista di letterature moderne e comparate* 23–4 (1970), 245–90.

Felski, Rita, *The Gender of Modernity* (Cambridge, MA: Harvard University Press, 1995).

Fink, Eugen, "The Oasis of Happiness: Towards an Ontology of Play", *Yale French Studies* 41 (1968), 19–30.

Fontanella, Luigi, *Storia di Bontempelli: tra i sofismi della ragione e le irruzioni dell'immaginazione* (Ravenna: Longo, 1997).

——, *Il surrealismo italiano* (Roma: Bulzoni, 1983).

Forti, Marco, "Romanzi straordinari" in: Caretti, Lucio, ed., *Palazzeschi oggi. Atti del convegno.* Firenze 6–8 novembre 1976 (Milano: Il Saggiatore, 1978), 125–30.

Frega, Donnalee, *Speaking in Hunger. Gender, Discourse and Consumption in "Clarissa"* (Columbia: University of South Carolina Press, 1998).

Frisby, David and Weatherstone, Mike (eds), *Simmel on Culture: Selected Writings* (Sage Publications Inc., 2000).

Furst, R. Lilian and Graham, W. Peter (eds), *Disorderly Eaters. Texts in Self-Empowerment* (University Park: The Pennsylvania State University Press, 1992).

Gargani, Aldo, *Il coraggio di essere* (Bari: Laterza, 1992).

Genette, Gerard, *Palimpsests. Literature in the Second Degree* (Lincoln and London: University of Nebraska Press, 1997).

Gentile, Emilio, "The Conquest of Modernity: From Modernist Nationalism to Fascism" *Modernism/Modernity* 1.3 (September 1994), 55–87.

Giacomoni, Silvia, Introduzione a *Nascita e morte della massaia* (Milano: La Tartaruga, 1982), 5–10.

Glielmo, Roselena, *La traversata dell'ironia: studi su Massimo Bontempelli* (Napoli: Guida, 1994).

Guglielmi, Guido, "The Paradoxes of Literary Modernity", The Yale Journal of Criticism 5 (1991), 217–33.

——, *La parola del testo: letteratura come storia* (Bologna: Il Mulino, 1993).

——, *La prosa italiana del novecento. Umorismo. Metafisica. Grottesco* (Torino: Einaudi, 1986).

——, *L'udienza del poeta. Saggi su Palazzeschi e il futurismo* (Torino: Einaudi, 1979).

Härmänmaa, Marja, "Beyond Anarchism Marinetti's Futurist (anti-)Utopia of Individualism and 'Artocracy'", *The European Legacy*, Vol. 14, n. 7 (2009), 857–71.

——, *Un patriota che sfidò la decadenza. F.T. Marinetti e l'idea dell'uomo nuovo fascista, 1929–1944* (Helsinki: Academia Scientiarum Fennicae, 2000).

Helstosky, Carol, "Recipe for the Nation: Reading Italian History through *La scienza in cucina* and *La cucina futurista*," *Food & Foodways* 11 (2003), 113–40.

Hewitt, Andrew, *Fascist Modernism: Aesthetics, Politics, and the Avant-Garde* (Stanford: Stanford University Press, 1993).

Heywood, Leslie, *Dedication to Hunger: the Anorexic Aesthetic in Modern Culture* (Berkeley: University of California Press, 1996).

Holland, Dorothy, "Bodies at Play: A General Economy of Performance", in Shannon Winnubst, ed., *Reading Bataille Now* (Bloomington and Indianapolis: Indiana University Press, 2007), 197–219.

Hollier, Dennis, *Against Architecture: The Writings of Georges Bataille* (Boston: MIT Press, 1992).

Huizinga, Johan, *Homo Ludens. A Study of the Play-Element in Culture* (Boston: The Beacon Press, 1950).

Jannini, Pasquale, *F.T. Marinetti. Scritti francesi* (Milano: Mondadori, 1983).

Kauffmann, Linda, *Discourses of Desire: Gender, Genre and Epistolary Fictions* (London and Ithaca: Cornell University Press, 1986).

Keeling, Kara K. and Scott T. Pollard (eds), *Critical Approaches to Food in Children's Literature* (New York and London: Routledge, 2008).

Kilgour, Maggie, *From Communion to Cannibalism: an Anatomy of Metaphors of Incorporation* (Princeton: Princeton University Press, 1990).

Kimball, A. Samuel, *The Infanticidal Logic of Evolution and Culture* (Newark: University of Delaware Press, 2007).

Klein, Julie R., "Nature's Metabolism: On Eating in Derrida, Agamben, and Spinoza" in *Research in Phenomenology*, 33 Koninklijke Brill NV, Leiden, The Netherlands (2003), 187–217.

Knapp, Bettina, "Jarry's The Supermale: The Sex Machine, the Food Machine, and the Bicycle Race. Is it a Question of Adaptation?" *Nineteenth-century French Studies* 18. 3–4 (1990), 492–507.

Krauss, Rosalind, *The Originality of the Avant-Garde and Other Modernist Myths* (Cambridge: The M.I.T. Press, 1985).

Labriola, Antonio, *Del materialismo storico* (Bari: Laterza, 1971).

Lanslots, Inge, "Un assaggio della narrativa italiana contemporanea: il sapore di chiusura" in Van den Bossche, Bart, Bastiaensen, Michel, Salvadori Lonergan, Corinna (eds), *Soavi sapori della cultura italiana. Atti del XIII Congresso dell' A.I.P.I.* Verona/Soave 27–29 agosto 1998 (Firenze: Cesati, 1999), 367–82.

Larizza, Mirella, *Fourier* (Firenze: Olschki, 2002).

Lepri, Laura, *Il funambolo incosciente. Aldo Palazzeschi 1905–1914* (Firenze: Olschki Editore, 1991).

Levenson, Michael, ed., *The Cambridge Companion to Modernism* (Cambridge: Cambridge University Press, 1999).

Lukacher, Maryline, *Maternal Fictions. Stendhal, Sand, Rachilde, and Bataille* (Durham and London: Duke University Press, 1994).

Luperini, Romano, "Laughter and Political Allegory in Pirandello", in Gianpaolo Biasin and Manuela Gieri, eds, *Luigi Pirandello. Contemporary Perspectives* (Toronto: University of Toronto Press, 1999), 107–24.

Madland, Helga Stipa, "Infanticide as Fiction: Goethe's Urfaust and Schiller's Kindsmörderin as Models." *The German Quarterly*, Vol. 62, n 1 (Winter 1989), 27–38.

Maffesoli, Michel, *The Shadow of Dionysus: a Contribution to the Sociology of the Orgy* (Albany: State University of New York Press, 1993).

Mahon, Alyce, "Hans Bellmer's Libidinal Politics" in Raymond Spiteri, Donald LaCoss, eds, *Surrealism, Politics and Culture* (Burlington: Ashgate, 2003).

Marin, Louis, *Food for Thought* (Baltimore and London: The Johns Hopkins University Press), 1989.

Marx, Karl, *The German Ideology* (Amherst, NY: Prometheus Books), 1998.

Mascia Galateria, Marinella, "L'autobiografia trasfigurata di Paola Masino" in *Avanguardia*, VI, 17, 110–31.

Maude, Ulrika, *Beckett, Technology and the Body* (Cambridge: Cambridge University Press, 2009).

Meuret, Isabelle, *L'anorexie créatrice* (Paris: Klincksieck, 2006).

——, *Writing Size Zero: Figuring Anorexia in Contemporary World Literatures* (Peter Lang, 2006).

Michaud, Eric, "The Real as a Primal Scene (A Note on Art and Devouring)" in *Utopia Post-Utopia. Configurations of nature and culture in recent sculpture and photography, January 29–March 27, 1988* (Boston: M.I.T. Press, 1988), 35–51.

Miller, Mary-Kay F., *(Re)Productions: Autobiography, Colonialism, and Infanticide* (New York: Peter Lang Publishing Incorporated, 2003).

Minerva, Nadia, "Il pudore della felicità: luoghi comuni e parole tabù nell'utopia francese tra Otto e Novecento", in Biondi, C., Imbroscio, C. Latil, M.J., Minerva, N., Pellandra, C., Sfragaro, A., Souberyan, B., Vecchi, P., eds, *La quête du bonheur et l'expression de la doleur dans la littérature et la pensée francaises* (Geneva: Droz, 1995), 181–8.

——, "Une utopie inquiète: Les *Lettres de Malaisie* de Paul Adam (1898)", *L'Esprit Créateur* 34, n. 4 (Winter 1994), 77–89.

Montanari, Massimo, *Food is Culture* (New York: Columbia University Press, 2006).

Morgan, E. Thaïs, ed., *Men Writing the Feminine. Literature, Theory, and the Question of Genders* (Albany: SUNY Press, 1994).

Moroni, Mario, *La presenza complessa. Identità e soggettività nelle poetiche del Novecento* (Ravenna: Longo, 1998).

Morpurgo Tagliabue, Guido, *Goethe e il romanzo* (Torino: Einaudi, 1991).

Morton, Timothy, ed., *Cultures of Taste/Theories of Appetite: Eating Romanticism* (New York: Palgrave Macmillan, 2004).

Mossmann, A. Carol, "Gastro-Exorcism: J.K. Huysmans and the Anatomy of Conversion", *Romanic Review*, 79 (1988), 341–53.

Muzzarelli, Giuseppina and Re, Lucia, Introduzione a *Il cibo e le donne nella cultura e nella storia. Prospettive interdisciplinary* (Bologna: Clueb, 2005), 7–27.

Novero, Cecilia, *Antidiets of the Avant-garde. From Futurist Cooking to Eat Art* (Minneapolis, London: University of Minnesota Press, 2010).

——, "Dada-Diets: Dysfunctional Physiologies of Devouring". *Seminar. A Journal of Germanic Studies*, Vol. XXVII, n. 1 (February 2001), 1–20.

O'Brien, Paul, *Mussolini in the First World War: the Journalist, the Soldier, the Fascist* (Oxford: Berg, 2004).

Onfray, Michel, *Le Ventre des philosophes. Critique de la raison diététique* (Paris: Bernard Grasset, 1989).

Pagano, Tullio, "Modernisms: From Bergson's Laughter to Pirandello's Humor", *The Italianist*, 17 (1997), 44–59.

Papini, Maria C., "Aldo Palazzeschi, Remy de Gourmont: Un gioco di *: riflessi*", *Rassegna della letteratura italiana*, 101.1 (January–April 1997), 116–30.

Pasi, Carlo, *Georges Bataille. La ferita dell'eccesso* (Torino: Bollati-Boringhieri, 2002).

Perniola, Mario, *Ritual Thinking: Sexuality, Death, World* (New York: Humanity Books, 2001).

——, *Il sex-appeal dell'inorganico* (Torino: Einaudi, 1994).

Pieri, Piero, *Ritratto del saltimbanco da giovane. Palazzeschi 1905–14* (Bologna: Patron, 1980).

Pinkus, Karen, *Bodily Regimes: Italian Advertising under Fascism* (Minneapolis: University of Minneapolis Press, 1995).

Plock, Vike Martina, "Modernism's Feast on Science: Nutrition and Diet in Joyce's *Ulysses*", *Literature and History*, Vol. 5, n. 2 (Autumn 2007), 30–42.

Pogliano, Claudio, "Scienza e stirpe: eugenica in Italia (1912–1939)," *Passato e presente* 5 (1984), 61–97.

Prado Bellei, Sérgio Luiz, "Brazilian Anthropophagy Revisited", in Francis Barker, Peter Hulme, Margaret Iversen, eds, *Cannibalism and the Colonial World* (Cambridge UK: Cambridge University Press, 1998), 87–109.

Radcliff-Umstead, Douglas, *The Mirror of Our Anguish: A Study of L. Pirandello's Narrative Writings* (Cranbury, NJ: Fairleigh Dickinson Univ. Press, 1978).

Re, Lucia, "Fame, cibo e antifascismo nella Massaia di Paola Masino" in Muzzarelli, Giuseppina and Re, Lucia, eds, *Il cibo e le donne nella cultura e nella storia. Prospettive interdisciplinari* (Bologna: Clueb, 2005), 165–81.

——, "Futurism and Fascism, 1914–1945" in Panizza, Letizia and Wood, Sharon, eds, *A History of Women's Writing in Italy* (Cambridge: Cambridge University Press, 2000), 190–204.

——, "Fascist Theories of 'Woman' and the Construction of Gender" in Pickering-Iazzi, Robin, ed., *Mothers of Invention. Women, Italian Fascism, and Culture* (Minneapolis, London: University of Minnesota Press, 1995).

Rigotti, Francesca, *La filosofia in cucina. Piccola critica della ragion culinaria* (Bologna: il Mulino, 1999).

Rorandelli, Tristana, "*Nascita e morte della massaia* di Paola Masino e la questione del corpo materno nel fascism," *Forum Italicum* 37, 1 (Spring 2003), 70–102.

Rousset, Jean, *Forme et signification. Essais sur les structures littéraires de Corneille à Claudel* (Paris: Librairie José Corti, 1962).

Saccone, Antonio, *Massimo Bontempelli: il mito del '900* (Napoli: Liguori, 1979).

Saccone, Edoardo, "Allegoria di Novembre: la sublimazione imperfetta di Aldo Palazzeschi", *MLN*, 92–1 (1977), 79–116.

Salaris, Claudia, *Cibo futurista: Dalla cucina nell'arte all'arte in cucina* (Roma: Stampa Alternativa, 2000).

Salibra, Elena, "Le acrobazie del tempo nel 'Purosangue' di Bontempelli" in Donati, Corrado ed., *Massimo Bontempelli, scrittore e intellettuale: atti del convegno*, Trento, 18–20 aprile 1991 (Editori Riuniti, 1992), 293–313.

Savoca, Giuseppe, *Concordanza delle poesie di Aldo Palazzeschi. Vol. I–II* (Firenze: Olschki, 1993).

Schaer, R., Claeys, G., Tower Sargent, L., eds, *Utopia* (New York, Oxford: Oxford University Press, 2000).

Schnapp, Jeffrey and de Castro Rocha, João Cezar, "Brazilian Velocities: On Marinetti's 1926 Trip to South America", *South Central Review*, vol. 13, n. 2/3 (Summer–Autumn, 1996), 105–56.

Sebaste, Beppe, *Lettere e filosofia. Poetica dell'epistolarità* (Firenze: Alinea Editrice, 1998).

Skubal, Susan, *Word of Mouth. Food and Fiction after Freud* (New York: Routledge, 2002).

Sica, Paola, "Il testo, il corpo e la cultura futurista: riflessioni sul romanzo *Un ventre di donna*." *Quaderni del '900*, 5 (2005), 11–23.

Somigli, Luca, *Legitimizing the Artist. Manifesto Writing and European Modernism, 1885–1915* (Toronto: University of Toronto Press, 2003).

——, "Modernism and the Quest for the Real: On Massimo Bontempelli's Minnie la candida" in Somigli, Luca and Moroni, Mario, eds, *Italian modernism: Italian culture between decadentism and avant-garde* (Toronto: University of Toronto Press, 2004), 309–50.

——, "On the Threshold: Space and Modernity in Marinetti's Early Manifestoes and tavole parolibere." *Rivista di studi italiani* 17, 1 (1999), 250–73.

Somigli, Luca and Capozzi, Rocco, eds, *Italian Prose Writers, 1900–1945* (Detroit, MI: Thomson Gale, 2002), 193–9.

Spackman, Barbara, *Fascist Virilities. Rhetoric, Ideology, and Social Fantasy in Italy* (Minneapolis: University of Minnesota Press, 1996).

Tamburri, Anthony, *Of Saltimbanchi and Incendiari. Aldo Palazzeschi and Avant-Gardism in Italy* (London and Toronto: Associated University Presses, 1990).

——, *A Reconsideration of Aldo Palazzeschi's Poetry (1905–1974)* (Lewiston, Queenston, Lampeter: The Edwin Mellen Press, 1998).

——, *Semiotics of Re-reading: Guido Gozzano, Aldo Palazzeschi, and Italo Calvino* (Madison, NJ: Fairleigh Dickinson University Press, 2003).

Taravacci, Piero, "Il realismo magico di Bontempelli." *Trimestre*, 13 (1980), 217–37.

Teal, Laurie, "The Hollow Women: Modernism, the Prostitute, and Commodity Aesthetics" in *Differences: A Journal of Feminist Cultural Studies* 7.3 (1995), 80–108.

Tempesti, Fernando, *Bontempelli* (Firenze: La Nuova Italia, 1974).

Terrile, Cristina, "La piramide di Aldo Palazzeschi: un'allegoria del discorso letterario", *The Italianist* 20 (2000), 98–120.

Ulmer, Gregory, "The Post-age", *Diacritics*, 11 (1981), 39–56.

Urgnani, Elena, *Sogni e visioni: Massimo Bontempelli fra surrealismo e futurismo* (Ravenna: Longo, 1991).

Urry, John, "The Consuming of Place" in Adam Jaworski and Annette Pritchard, eds, *Discourse, Communication and Tourism* (Clevedon and Buffalo: Channel View Publications, 2005), 19–27.

Valesio, Paolo, "Spiritualità alla svolta del secolo: un anti-apologo", *Annali di storia dell'esegesi*, 10/1 (1993), 145–63.

Van den Bossche, Bart, "Il cibo nella narrativa del Novecento: appunti per una tipologia" in Van den Bossche, Bart, Bastiaensen, Michel, Salvadori Lonergan, Corinna, eds, *Soavi sapori della cultura italiana. Atti del XIII Congresso dell' A.I.P.I.* Verona/Soave 27–29 agosto 1998 (Firenze: Cesati, 1999), 480–92.

Venè, Gianfranco, *Il capitale e il poeta. Storia dei rapporti tra il capitalismo e la letteratura italiana dall'Illuminismo a Pirandello* (Milano: Sugar, 1972).

Verdicchio, Massimo, "Reading Perniola Reading. An Introduction." *Ritual Thinking: Sexuality, Death, World. Mario Perniola* (New York: Humanity Books, 2001), 15–41.

Vice, Sue, *Introducing Bakhtin* (Manchester and New York: Manchester University Press, 1997).

Index

ITALIAN MODERNITIES

Edited by
Pierpaolo Antonello and Robert Gordon,
University of Cambridge

The series aims to publish innovative research on the written, material and visual cultures and intellectual history of modern Italy, from the 19th century to the present day. It is open to a wide variety of different approaches and methodologies, disciplines and interdisciplinary fields: from literary criticism and comparative literature to archival history, from cultural studies to material culture, from film and media studies to art history. It is especially interested in work which articulates aspects of Italy's particular, and in many respects, peculiar, interactions with notions of modernity and postmodernity, broadly understood. It also aims to encourage critical dialogue between new developments in scholarship in Italy and in the English-speaking world.
Proposals are welcome for either single-author monographs or edited collections (in English and/or Italian). Please provide a detailed outline, a sample chapter, and a CV. For further information, contact the series editors, Pierpaolo Antonello (paa25@cam.ac.uk) and Robert Gordon (rscg1@cam.ac.uk).

Vol. 1 Olivia Santovetti: *Digression: A Narrative Strategy in the Italian Novel*. 260 pages, 2007.
ISBN 978-3-03910-550-2

Vol. 2 Julie Dashwood and Margherita Ganeri (eds): *The Risorgimento of Federico De Roberto*. 339 pages, 2009.
ISBN 978-3-03911-858-8

Vol. 3 Pierluigi Barrotta and Laura Lepschy with Emma Bond (eds): *Freud and Italian Culture*. 252 pages, 2009.
ISBN 978-3-03911-847-2

Vol. 4 Pierpaolo Antonello and Florian Mussgnug (eds): *Postmodern Impegno: Ethics and Commitment in Contemporary Italian Culture*. 354 pages, 2009.
ISBN 978-3-0343-0125-1

Vol. 5 Florian Mussgnug: *The Eloquence of Ghosts: Giorgio Manganelli and the Afterlife of the Avant-Garde*. 257 pages, 2010.
ISBN 978-3-03911-835-9

Vol. 6 Christopher Rundle: *Publishing Translations in Fascist Italy*. 268 pages, 2010.
ISBN 978-3-03911-831-1

Vol. 7 Jacqueline Andall and Derek Duncan (eds): *National Belongings: Hybridity in Italian Colonial and Postcolonial Cultures*. 251 pages, 2010.
ISBN 978-3-03911-965-3

Vol. 8 Emiliano Perra: *Conflicts of Memory: The Reception of Holocaust Films and TV Programmes in Italy, 1945 to the Present*. 299 pages, 2010.
ISBN 978-3-03911-880-9

Vol. 9 Alan O'Leary: *Tragedia all'italiana: Italian Cinema and Italian Terrorisms, 1970–2010*. 300 pages, 2011.
ISBN 978-3-03911-574-7

Vol. 10 Robert Lumley: *Entering the Frame: Cinema and History in the Films of Yervant Gianikian and Angela Ricci Lucchi*. 228 pages. 2011.
ISBN 978-3-0343-0113-8

Vol. 11 Enrica Maria Ferrara: *Calvino e il teatro: storia di una passione rimossa*. 284 pages, 2011.
ISBN 978-3-0343-0176-3

Vol. 12 Niamh Cullen: *Piero Gobetti's Turin: Modernity, Myth and Memory*. 343 pages, 2011.
ISBN 978-3-0343-0262-3

Vol. 13 Jeffrey T. Schnapp: *Modernitalia.* 338 pages, 2012.
ISBN 978-3-0343-0762-8

Vol. 14 Eleanor Canright Chiari: *Undoing Time: The Cultural Memory of an Italian Prison*. 275 pages, 2012.
ISBN 978-3-0343-0256-2

Vol. 15 Alvise Sforza Tarabochia: *Psychiatry, Subjectivity, Community: Franco Basaglia and Biopolitics*. 226 pages, 2013.
ISBN 978-3-0343-0893-9

Vol. 16 Katharine Mitchell and Helena Sanson (eds): *Women and Gender in Post-Unification Italy: Between Private and Public Spheres*. 282 pages, 2013.
ISBN 978-3-0343-0996-7

Vol. 17 Enrico Cesaretti: *Fictions of Appetite: Alimentary Discourses in Italian Modernist Literature.* 280 pages, 2013.
ISBN 978-30343-0971-4